Beginning Algebra II

Michael Gaul, Barbara Goldner, Edgar Jasso, Deanna Li,
Pam Lippert, Eileen Murphy, Sam Wilson

3rd Edition
Fall 2018

Contents

6   Factoring 7

   6.1   **Greatest Common Factor and Grouping**   **7**
      Factoring the Greatest Common Factor 7
      Factoring by Grouping 12

   6.2   **Factoring trinomials - Part I**   **16**
      Perfect Square Trinomial 20
      Difference of Two Squares 21
      Greatest Common Factor 23

   6.3   **Factoring trinomials - Part II**   **28**
      List and Check Method 28
      AC-Method 34

   6.4   **Factoring Special Products**   **40**
      Difference of Two Squares 40
      Perfect Square Trinomial 44

   6.5   **Factoring Strategies**   **49**

   6.6   **Solving Quadratic Equations by Factoring**   **54**

7   Radicals Part I 65

   7.1   **Square Roots**   **65**
      Evaluating Square Roots 66
      Real Numbers 68
      Approximating Irrational Numbers 69

Expressions with Square Roots 70

7.2   Simplifying square roots   73
Squaring a Square Root Term 76

8   Graphs of Quadratic Equations 81

8.1   Square Root Property   81
Square Root Property 81
Quadratic Equations 83

8.2   Completing the square   96
Forming a Perfect Square Trinomial Algebraically 96
Solving Quadratic Equations by Completing the Square 106

8.3   The Quadratic Formula   117
Quadratic Formula Derivation 117

8.4   Strategies for solving Quadratic Equations   125

8.5   Graphs of Quadratic Equations   134
Properties of the Graph of a Quadratic Equation 137
Vertex and Line of Symmetry 137

8.6   Applications of quadratic Part II   159
Pythagorean Theorem 162
Maximum-Minimum Problems 166

9   Radicals Part II 175

9.1   Simplifying Radicals   175
Simplifying Radicals Involving Variables 181

9.2   Adding and Subtracting Radicals   184
Like and Unlike Radical Expressions 184
Adding and Subtracting Radical Expressions 185

9.3   Multiplying and Dividing Radicals   189
Multiplication of Radicals 189
Division of Radicals 194
Rationalizing the Denominator 197

9.4   Square Root Equations   201
Application Problems 207

10   Rational Expressions 215

10.1   Introduction to Rational Expressions   215
Common Denominators 216
Determining When a Rational Expression is Undefined 218
Simplifying Rational Expressions 219

10.2 Multiplying and Dividing Rational Expressions   **230**

     Multiplication of Fractions 230
     Division of Fractions 230
     Multiplication and Division of Rational Expressions 232

10.3 Adding and Subtracting Rational Expressions   **239**

     Common Denominators 239
     Unlike Denominators 241

10.4 Complex Fractions   **250**

     Complex Fractions 250
     Complex Rational Expressions 255

10.5 Solving Rational Equations   **262**

     Answer Key 286

Preface

Mathematics instructors from North Seattle College (NSC) created this workbook to better serve students and to align the curriculum with NSC learning outcomes. This workbook assumes no prior knowledge of algebra, but assumes that students are familiar with the basic rules of arithmetic. This is the second part of a series of two workbooks: Beginning Algebra I and Beginning Algebra II. After mastering the topics included in this series students will be prepared for a course in Intermediate Algebra.

This workbook is organized as follows.

- **Chapters**

  - ◇ **Sections** are the main instructional component for each chapter. This is where ideas are introduced.

  - ◇ **Worked Examples** provide further explanation of a concept. It is recommended that students read and work through these examples carefully.

  - ◇ **Class Examples** provide classroom examples for instructors as concepts are being developed.

  - ◇ **You Try** are problems embedded in the workbook to help reinforce the concepts learned. It is recommended that students work through the problems in the order they appear, showing as much work as possible in a neat and organized fashion. Space for work is provided in the workbook. However, there may not be enough space so it is recommended that a notebook be used for all math work.

- **Section Exercises**

  - ◇ Each section ends with a set of **exercises.** The only way to learn math is by practice. We suggest that students attempt the exercises on their own first before seeking help. There is no space to show work for the exercises. It is highly recommended that students keep an organized notebook for the class. Students should do the exercises in their notebook, showing all worked solutions in a neat and organized way, so they can refer to them easily.

◇ Solutions to the Exercises are at the end of the workbook. It is very important to verify that you have the correct answer before proceeding to the next problem.

6. Factoring

6.1 Greatest Common Factor and Grouping

Objective: To factor polynomials by finding the greatest common factor and by grouping

Factoring is the reverse process of multiplication. Let us review multiplying a monomial by a binomial. Remember that the distributive property tells us to

> **Example 1** Multiply $3x(5x+8)$
>
> **Solution.**
>
> $$3x(5x+8) = 3x(5x) + 3x(8)$$
> $$= 15x^2 + 24x \qquad\qquad \text{Our solution}$$

Factoring the Greatest Common Factor

To **factor** an expression is to rewrite the expression as a product. For integers, we usually express them as a product of prime numbers. For example, 6 can be rewritten as a product of prime numbers, 2 and 3, that is, $6 = 2 \cdot 3$. Similarly, to factor a polynomial means to rewrite the polynomial as a product.

To factor out the greatest common factor (GCF) of an algebraic expression, ask yourself: What are the greatest factors that are in common for each term?

Example 2 Identify the greatest common factor (GCF) and completely factor the expression, $5a^3 + 15a$.

Solution.

$$\text{Factors of } 5a^3: 5 \cdot a \cdot a \cdot a \qquad\qquad \text{Factors of } 15a: 3 \cdot 5 \cdot a$$

Both terms, $5a^3$ and $15a$, have a greatest common factor $5 \cdot a = 5a$. Let us now factor the expression.

$$\begin{aligned}
5a^3 + 15a &= 5 \cdot a \cdot a \cdot a \cdot + 3 \cdot 5 \cdot a \\
&= 5 \cdot a(a \cdot a + 3) \\
&= 5a(a^2 + 3)
\end{aligned}$$

Check to make sure the expression is factored correctly by applying the distributive property.

$$5a(a^2 + 3) = 5a^3 + 15a \quad \checkmark$$

Exercise 1 Class Example

Identify the greatest common factor (GCF) and factor each expression completely. Be sure to check your answer.

a) $8h + 24$

c) $4x^2 - 12x^3 + 16x^5$

b) $3y^2 - 12y$

d) $5ab^2 + 10a^2b^2 + 15a^2b$

Exercise 2 You Try
Identify the greatest common factor (GCF) and factor each expression completely. Be sure to check your answer.

a) $15m + 40$

c) $2a^3 + 14a^2 + 7a$

b) $14y^5 - 49y^2$

d) $15mn + 12m - 3mn^2$

World View Note The first recorded algorithm for finding the greatest common factor comes from the Greek mathematician Euclid, around the year 300 BC!

When a polynomial has a negative leading coefficient, we usually factor out a negative sign, making the GCF a negative number.

Example 3 Factor $-4a + 12b - 8c$ completely.

Solution.
Since the leading coefficient of the given polynomial is negative and 4 is a common factor, the given expression's GCF is -4. Factoring -4 out, we get the following.

$$-4a + 12b - 8c = -4(a - 3b + 2c)$$

Notice the signs of the polynomial inside the parenthesis are all the opposite signs from the original polynomial. Lets check that we have the correct factors by applying the distributive property.

$$-4(a - 3b + 2c) = -4a + 12b - 8c \checkmark$$

Example 4 Factor $-x^2 + 5x - 2$ completely.

Solution. Since the leading coefficient of the given polynomial is negative and there are no other common factors, the GCF is -1. The polynomial factors as follows.

$$-x^2 + 5x - 2 = -1(x^2 - 5x + 2)$$
$$= -(x^2 - 5x + 2)$$

Check to verify that factor is correct.

$$-(x^2 - 5x + 2) = -x^2 + 5x - 2 \checkmark$$

Exercise 3 Class Example
Identify the greatest common factor (GCF) and factor each expression completely. Be sure to check your answer.

a) $-x - 2y + 3$

b) $-18p^3 + 12p^2 - 6p$

Exercise 4 You Try
Identify the greatest common factor (GCF) and factor each expression completely. Be sure to check your answer.

a) $-15x^3 + 9x^2$

b) $-6xyz + 12xz - 15xy$

The greatest common factor is not always a monomial. Sometimes, it can be a binomial or any other polynomial.

Example 5 Factor $4x(x - 3) + 7(x - 3)$ completely.

Solution.
Let us take a look at the factors of each term.

Factors of the first term, $4x(x - 3)$: $4 \cdot x \cdot (x - 3)$
Factors of the second term, $7(x - 3)$: $7 \cdot (x - 3)$

Both terms have a greatest common factor $(x-3)$. Factoring $(x-3)$ out, we get the following.

$$4x(x-3)+7(x-3) = (x-3)(4x+7)$$

Check to verify the factors are correct. Applying the distributive property to the factored expression, we get the following.

$$(x-3)(4x+7) = x(4x+7)-3(4x+7)$$
$$= 4x^2+7x-12x-21$$
$$= 4x^2-5x-21$$

Applying the distributive property to the given expression, we get the following.

$$4x(x-3)+7(x-3) = 4x^2-12x+7x-21$$
$$= 4x^2-5x-21$$

The factored form and the given expression yield the same result, $4x^2-5x-21$. Therefore, $(x-3)(4x-7)$ must be the correct factor for the given expression.

Exercise 5 Class Example
Identify the greatest common factor (GCF) and factor each expression completely. Be sure to check your answer.

a) $a^2(a+5)+9(a+5)$

b) $10x^2(2x-3)-15x(2x-3)$

Exercise 6 **You Try**
Identify the greatest common factor (GCF) and factor each expression completely. Be sure to check your answer.

  a) $a(a+7)+5(a+7)$ b) $8(t-3)-3t(t-3)$

Factoring by Grouping

Sometimes there are no common factors to all the terms of the polynomial but there may be factors common to some of the terms. Let us see how the grouping technique helps to factor expressions.

Example 6 Factor $10ab+15b+4a+6$ completely.

Solution.

The expression has no common factor to all of its terms. Let us use the grouping technique to factor the first two terms and the last two terms.

$$\underbrace{10ab+15b}+\underbrace{4a+6} \qquad\qquad \text{Find GCF for each group}$$

$$= 5b(2a+3)+2(2a+3) \qquad\qquad \text{Factor out GCF, } (2a+3)$$

$$= (2a+3)(5b+2) \qquad\qquad \text{Our Solution}$$

Check that factors are correct.

$$(2a+3)(5b+2) = 2a(5b+2)+3(5b+2)$$
$$= 10ab+4a+15b+6$$
$$= 10ab+15b+4a+6 \ \checkmark$$

Example 7 Factor $5xy - 8x - 10y + 16$ completely.

Solution.

The expression has no common factor to all of its terms. Let us factor by grouping the first two terms and the last two terms. Note that the leading coefficient of the third term is negative. Its GCF must be a negative number.

$$\underbrace{5xy - 8x}\ \underbrace{-10y + 16} \qquad \text{Find GCF for each group}$$
$$= x(5y - 8) - 2(5y - 8) \qquad \text{Factor out GCF, } (5y - 8)$$
$$= (5y - 8)(x - 2) \qquad \text{Our Solution}$$

Check that factors are correct.

$$(5y - 8)(x - 2) = 5y(x - 2) - 8(x - 2)$$
$$= 5xy - 10y - 8x + 16$$
$$= 5xy - 8x - 10y + 16 \ \checkmark$$

Example 8 Factor $6x^3 - 15x^2 + 2x - 5$ completely.

Solution.

The expression has no common factor to all of its terms. Let us factor by grouping the first two terms and the last two terms. Notice that the last two terms have no common factor. In this case, we will factor out a 1.

$$\underbrace{6x^3 - 15x^2} + \underbrace{2x - 5} \qquad \text{Find GCF for each group}$$
$$= 3x^2(2x - 5) + 1(2x - 5) \qquad \text{Factor out GCF, } (2x - 5)$$
$$= (2x - 5)(3x^2 + 1) \qquad \text{Our Solution}$$

Check that factors are correct.

$$(2x - 5)(3x^2 + 1) = 2x(3x^2 + 1) - 5(3x^2 + 1)$$
$$= 6x^3 + 2x - 15x^2 - 5$$
$$= 6x^3 - 15x^2 + 2x - 5 \ \checkmark$$

Exercise 7 Class Example
Factor each expression completely. Be sure to check your answer.

a) $xy + 2x + 4y + 8$

c) $49y^2 - 14y - 14y + 4$

b) $ab - 3b + 7a - 21$

d) $2b^3 - 6b^2 + b - 3$

Exercise 8 You Try
Factor each expression completely. Be sure to check your answer.

a) $xy + 2y + 6x + 12$

c) $w^3 - 5w^2 - w + 5$

b) $3ax + 21x - a - 7$

d) $12y^2 - 21y + 20y - 35$

6.1: Exercises

Factor each expression completely.

1. $45x^2 - 25$

2. $56 - 35p$

3. $50x - 80y$

4. $7ab - 35a^2b$

5. $-3a^2b + 6a^3b^2$

6. $-32n^3 + 32n^2 + 8n$

7. $-5x^2 + 25x^3 - 15x^4$

8. $21p^2 + 30p + 27$

9. $-10x^4 + 20x^2 + 12x$

10. $30b^2 + 5ab - 15a^2$

11. $-27x^2y^2 + 12xy^2 - 9y^2$

12. $3x^3 + 15x^2 + 2x + 10$

13. $3n^3 - 2n^2 - 9n + 6$

14. $40r^3 - 8r^2 - 25r + 5$

15. $15b^3 + 21b^2 - 35b - 49$

16. $7xy - 49x + 5y - 35$

17. $16xy - 56x + 2y - 7$

18. $7n^3 + 21n^2 - 5n - 15$

19. $3mn - 8m + 15n - 40$

20. $8xy + 56x - y - 7$

21. $28p^3 + 21p^2 + 20p + 15$

22. $14v^3 + 10v^2 - 7v - 5$

23. $4y^3 - 12y^2 + 8y - 24$

24. $30x^3 - 20x^2 - 6x + 4$

25. $14u^2 + 42u + 4uv + 12v$

6.2   Factoring trinomials - Part I

Objective: To factor trinomials where the leading coefficient is one

Over the next two sections, we will learn how to factor trinomials of the form $ax^2 + bx + c$. This type of trinomial is known as a **quadratic** and it has a degree of 2.

- ax^2 is known as the **quadratic term** and a is the leading coefficient

- bx is referred to as the **linear term**

- c is the **constant term**

In this section, we will focus on trinomials where the leading coefficient is one. In the next section, we will cover quadratics where $a \neq 1$, after factoring the greatest common factor.

Since factoring is the reverse process of multiplication, let us review binomial multiplication. Let us multiply $(x+6)(x+4)$.

$$
\begin{aligned}
(x+6)(x+4) &= x(x+4) + 6(x+4) && \text{Distribute each monomial} \\
&= x^2 + 4x + 6x + 24 && \text{Combine like terms} \\
&= x^2 + 10x + 24 && \text{Our Solution}
\end{aligned}
$$

There are some things to notice about our answer.

1. The answer to the product of our 2 binomials is a trinomial. Therefore, if a trinomial can be factored, the factors will be two binomials.

2. x^2 is the product of the first terms of each binomial.

3. b is the sum of the constants found in each binomial. From the above example, $10 = 4 + 6$.

4. c is the product of the constants. From the above example, $24 = 6(4)$.

Consequently, in order to factor a trinomial, we need to find factors of c that sum to b.

Example 1 Factor $x^2 + 8x + 15$ completely.

Solution.

Let us get started with things we know. If this trinomial is factorable, the answer will be the product of two binomials. Since x^2 can only factor as $x \cdot x$, we can write

$$
x^2 + 8x + 15 = (x \qquad)(x \qquad)
$$

Next, we need to find factors of 15 that sum to 8. Since 8 and 15 are both positive, we only consider positive factors.

Positive Factors of 15	Product of Factors	Sum of Factors
1, 15	$1 \cdot 15 = 15$	$1 + 15 = 16$
3, 5	$3 \cdot 5 = 15$	$3 + 5 = 8$

The last pair gives us what we need. We put those values into the binomials to get our final answer.

$$x^2 + 8x + 15 = (x+3)(x+5)$$

To check that we have the correct factors, we perform binomial multiplication and confirm we get back the original trinomial.

$$\begin{aligned}(x+3)(x+5) &= x(x+5) + 3(x+5) & \text{Distribute each monomial} \\ &= x^2 + 5x + 3x + 15 & \text{Combine like terms} \\ &= x^2 + 8x + 15 \ \checkmark \end{aligned}$$

Example 2 Factor $x^2 + 8x - 20$ completely.

Solution.
If this trinomial is factorable, the answer will be the product of two binomials. Since x^2 can only factor as $x \cdot x$, we can write

$$x^2 + 8x - 20 = (x \qquad)(x \qquad)$$

Next, we need to find factors of -20 that sum to 8. Given that the product is negative, we have to consider combinations of positive and negative factors.

Factors of -20	Product of Factors	Sum of Factors
$1, -20$	$1 \cdot -20 = -20$	$1 + (-20) = -19$
$2, -10$	$2 \cdot -10 = -20$	$2 + (-10) = -8$

Notice that the last pair comes very close to what we want but the sign of the sum is the opposite of what we want. This means that we need to swap the signs of the factors. Instead of 2 and -10, use -2 and 10. We see that the product of the factors, $-2 \cdot 10 = -20$ and the sum of the factors, $-2 + 10 = 8$ are the combinations we want.

$$x^2 + 8x - 20 = (x-2)(x+10)$$

Confirm that we have the correct factors by performing binomial multiplication.

$$\begin{aligned}(x-2)(x+10) &= x(x+10) - 2(x+10) & \text{Distribute each monomial} \\ &= x^2 + 10x - 2x - 20 & \text{Combine like terms} \\ &= x^2 + 8x - 20 \ \checkmark \end{aligned}$$

Example 3 Factor $x^2 - 8x + 9$ completely.

Solution.

If this trinomial is factorable, the answer will be the product of two binomials. Since x^2 can only factor as $x \cdot x$, we can write

$$x^2 - 8x + 9 = (x \qquad)(x \qquad)$$

Next, we need to find factors of 9 that sum to -8. Given that the product is positive with a negative sum, we will only consider negative factors of 9.

Negative Factors of 9	Product of Factors	Sum of Factors
$-1, -9$	$(-1) \cdot (-9) = (-9)$	$(-1) + (-9) = -10$
$-3, -3$	$(-3) \cdot (-3) = 9$	$(-3) + (-3) = -6$

We have exhausted all the negative factors of 9 and none of them sum up to a -8. This trinomial is not factorable. A polynomial that is not factorable is called **prime**. Therefore, we say that $x^2 - 8x + 9$ is prime.

Exercise 1 **Class Example**

Factor the following trinomials completely, if possible. If the trinomial is not factorable, identify it as prime. Be sure to check your answer.

a) $p^2 + 5p + 6$ c) $y^2 + 9y - 10$

b) $x^2 - 4x + 3$ d) $m^2 + 4m + 14$

Exercise 2 You Try

Factor the following trinomials completely, if possible. If the trinomial is not factorable, identify it as prime. Be sure to check your answer.

a) $a^2 + 9a + 18$ e) $c^2 - 10c + 2$

b) $y^2 - 7y + 6$ f) $p^2 + 13p + 36$

c) $y^2 - 7y - 18$ g) $m^2 + 2m - 3$

d) $x^2 - 4x - 12$ h) $n^2 - 5n - 6$

Let us summarize what we see happening with factoring of trinomials when the leading coefficient is 1.

1. When the constant term of the trinomial is positive, its factors must have the same signs because two positive factors yield a positive product. Similarly, two negative factors yield a positive product.

    -   If the linear term is positive, the trinomial factors as $(x + \underline{})(x + \underline{})$. For example, $x^2 + 7x + 10 = (x+5)(x+2)$.
    -   If the linear term is negative, the trinomial factors as $(x - \underline{})(x - \underline{})$. For example, $x^2 - 7x + 10 = (x-5)(x-2)$.

2. When the constant term of the trinomial is negative, its factors must have different signs because a positive and a negative factor yield a negative product. In this case the trinomial factors as $(x + \underline{})(x - \underline{})$. The sign of the linear term will indicate which factor is positive and which is negative. Let us take a look at some examples.

$$x^2 + 3x - 10 = (x+5)(x-2)$$
$$x^2 - 3x - 10 = (x-5)(x+2)$$

Perfect Square Trinomial

Example 4 Factor $x^2 - 10x + 25$ completely.

Solution.

If this trinomial is factorable, the answer will be the product of two binomials. Since x^2 can only factor as $x \cdot x$, we can write

$$x^2 - 10x + 25 = (x \qquad)(x \qquad)$$

Next, we need to find factors of 25 that sum to -10. Since the constant term, 25, is positive and the linear term has a negative coefficient, we will only consider negative factors.

Negative Factors of 25	Product of Factors	Sum of Factors
$-1, -25$	$(-1) \cdot (-25) = 25$	$(-1) + (-25) = -26$
$-5, -5$	$(-5) \cdot (-5) = 25$	$(-5) + (-5) = -10$

The last pair gives us what we need. We put those values into the binomials to get our final answer.

$$x^2 - 10x + 25 = (x-5)(x-5)$$
$$= (x-5)^2$$

To check that we have the correct factors, we perform binomial multiplication and confirm we get back the original trinomial.

$$(x-5)^2 = (x-5)(x-5)$$
$$= x(x-5)-5(x-5)$$
$$= x^2-5x-5x+25$$
$$= x^2-10x+25 \checkmark$$

Notice that this trinomial factors as two identical binomials. We can rewrite the trinomial as the binomial squared. Trinomials that factor as the square of a binomial are called **Perfect Square Trinomials**. We will explore more of these trinomials in a later section.

Exercise 3 **Class Example**
Factor the following trinomials completely. Be sure to check your answer.

a) p^2+6p+9 b) $x^2-12x+36$

Exercise 4 **You Try**
Factor the following trinomials completely. Be sure to check your answer.

a) h^2+2h+1 b) $y^2-14y+49$

Difference of Two Squares

Example 5 Factor x^2-25 completely.

Solution.
You might notice that this is not a trinomial. It is missing the linear term. However, we can "pencil" in the middle term with a coefficient of 0. (Remember that adding 0 to an expression does not change its value.)

Consider then $x^2 + \mathbf{0}x - 25$. We need to find factors of -25 that sum to 0. Since the constant term is negative, we will only consider a positive and a negative factor.

Factors of -25	Product of Factors	Sum of Factors
$-1, 25$	$(-1) \cdot (25) = -25$	$(-1) + (25) = 24$
$1, -25$	$(1) \cdot (-25) = -25$	$(1) + (-25) = -24$
$-5, 5$	$(-5) \cdot (5) = -25$	$(-5) + (5) = 0$
$5, -5$	$(5) \cdot (-5) = -25$	$(5) + (-5) = 0$

The last two pairs give us what we need. We put those values into the binomials to get our final answer.

$$x^2 - 25 = (x - 5)(x + 5)$$
$$x^2 - 25 = (x + 5)(x - 5)$$

Either answer is correct because multiplication is commutative. To check that we have the correct factors, we perform binomial multiplication and confirm we get back the original trinomial.

$$\begin{aligned}(x - 5)(x + 5) &= x(x + 5) - 5(x + 5) \\ &= x^2 + 5x - 5x - 25 \\ &= x^2 - 25 \ \checkmark\end{aligned}$$

$$\begin{aligned}(x + 5)(x - 5) &= x(x - 5) + 5(x - 5) \\ &= x^2 - 5x + 5x - 25 \\ &= x^2 - 25 \ \checkmark\end{aligned}$$

We can only find factors of the constant term that sum to 0 when the constant term is a negative perfect square. Consequently, when a binomial is of the form $a^2 - b^2$, it is called the **Difference of Two Squares** and factors as follows:

$$a^2 - b^2 = (a + b)(a - b)$$

We will go more in depth on this topic in a later section.

Exercise 5 **Class Example**
Factor the following completely, if possible. If it is not factorable, identify it as prime. Be sure to check your answer.

a) $p^2 - 49$ b) $y^2 - 100$

Exercise 6 You Try
Factor the following completely, if possible. If it is not factorable, identify it as prime. Be sure to check your answer.

a) $x^2 - 36$ b) $c^2 - 81$

Greatest Common Factor

When factoring any polynomial, you always want to first factor out the Greatest Common Factor (GCF).

Example 6 Factor $5x^2 - 25x - 30$ completely.

Solution.
Factor out the GCF first.

$$5x^2 - 25x - 30 = 5(x^2 - 5x - 6) \qquad \text{Factor out GCF 5}$$

Now we try to factor the trinomial inside the parenthesis. We want to find factors of -6 that sum to -5. The factors of -6 that will sum to -5 are -6 and 1. The trinomial factors completely as follows.

$$5x^2 - 25x - 30 = 5(x^2 - 5x - 6)$$
$$= 5(x - 6)(x + 1)$$

Check that we have the correct factors.

$$5(x - 6)(x + 1) = (5x - 30)(x + 1)$$
$$= 5x(x + 1) - 30(x + 1)$$
$$= 5x^2 + 5x - 30x - 30$$
$$= 5x^2 - 25x - 30 \ \checkmark$$

Exercise 7 Class Example

Factor the following completely, if possible. If it is not factorable, identify it as prime. Be sure to check your answer.

a) $3y^2 + 24y + 48$

c) $5m^2 - 20$

b) $2x^3 - 10x^2 - 12x$

d) $3t^2 - 21t + 36$

Exercise 8 You Try

Factor the following completely, if possible. If it is not factorable, identify it as prime. Be sure to check your answer.

a) $2x^2 - 2$

c) $4x^2 + 8x - 32$

b) $c^3 - 6c^2 + 9c$

d) $10p^3 + 50p^2 + 60p$

Exercise 9 **Class Example**

a) Given the quadratic and constant terms of a trinomial, find four possible options, 2 positive and 2 negative values, for the linear term: $x^2 + \underline{\quad} + 12$. Factor each trinomial.

b) Given the quadratic and linear terms of a trinomial, find four possible options, 2 positive and 2 negative values, for the constant term: $x^2 - 8x\underline{\quad}$. Factor each trinomial.

Exercise 10 You Try

a) Given the quadratic and constant terms of a trinomial, find four possible options, 2 positive and 2 negative values, for the linear term: $x^2 + \_\_ + 18$. Factor each trinomial.

b) Given the quadratic and linear terms of a trinomial, find four possible options, 2 positive and 2 negative values, for the constant term: $x^2 + 7x + \_\_$. Factor each trinomial.

6.2: Exercises

Factor each expression completely.

1. $p^2 + 8p + 12$

2. $x^2 - x - 12$

3. $n^2 - 9n + 8$

4. $x^2 + x - 30$

5. $x^2 - 9x - 10$

6. $b^2 + 12b + 32$

7. $y^2 + 3y + 4$

8. $b^2 - 17b + 70$

9. $x^2 + 3x - 70$

10. $x^2 + 3x - 18$

11. $n^2 - 8n + 15$

12. $a^2 - 6a - 27$

13. $p^2 + 15p + 54$

14. $p^2 + 7p - 30$

15. $n^2 - 15n + 56$

16. $x^2 - 9$

17. $m^2 - 49$

18. $p^2 - 1$

19. $x^2 - 2x + 1$

20. $m^2 + 12x + 36$

21. $y^2 + 16y + 64$

22. $n^2 + 2n + 4$

23. $3v^2 - 12v + 18$

24. $-x^2 + 18x - 81$

25. $6x^2 + 18x + 12$

26. $4x^2 + 20x + 24$

27. $-5n^2 + 45n - 40$

28. $5v^3 + 20v^2 - 25v$

29. $4n^2 - 64$

30. $y^3 - 25y$

6.3   Factoring trinomials - Part II

Objective: To factor trinomials where the leading coefficient is not equal to one

Since factoring is the reverse process of multiplication, let us review what happens when you multiply two binomials like $(2x+3)(3x+1)$.

$$(2x+3)(3x+1) = 2x(3x+1)+3(3x+1)$$
$$= 6x^2+2x+9x+3$$
$$= 6x^2+11x+3$$

We notice the following from the above product.

- We obtained the quadratic term, $6x^2$ when we multiplied the leading terms of our binomial, that is, $2x \cdot 3x = 6x^2$.

- We obtained the constant term, 3, when we multiplied the second terms of our binomials, that is, $3 \cdot 1 = 3$.

- We obtained the linear term, $11x$, in a more complicated way. It is the sum of the two terms, $2x+9x = 11x$.

List and Check Method

The strategy we will use to factor a trinomial will focus on the factors of the first and last terms of the trinomial. This could give us several options. We will check to see which of the factors are the right ones.

Example 1 Factor $3x^2+7x+2$ completely.

Solution.
Start by obtaining the factors of the quadratic term, $3x^2$. There is only one possibility: $3x$ and x. So the binomial factors we are looking for will have the following form.

$$(3x\quad)(x\quad)$$

Next, obtain the factors of the constant term, 2. There are two possibilities. They are 2 and 1 or -2 and -1. Both factors will yield a product of 2. However, the linear term is positive, so only positive factors will work. This gives us a total of 2 possibilities for the factor of the given trinomial.

$$(3x+2)(x+1) \quad \text{or} \quad (3x+1)(x+2)$$

To check which one is correct, we multiply the factors.

$$(3x+2)(x+1) = 3x(x+1)+2(x+1)$$
$$= 3x^2+3x+2x+2$$
$$= 3x^2+5x+2 \quad \times$$

$$(3x+1)(x+2) = 3x(x+2)+1(x+2)$$
$$= 3x^2+6x+x+2$$
$$= 3x^2+7x+2 \quad \checkmark$$

Therefore, $3x^2+7x+2 = (3x+1)(x+2)$.

The method described above is known as the *list and check* method (or *guess and check method*). With practice, you will become more efficient with this method, at least when the numbers involved are simple enough.

Exercise 1 **Class Example**
Factor $3x^2+8x+5$ completely. Be sure to check your answer.

Exercise 2 **You Try**
Factor $2x^2+11x+5$ completely. Be sure to check your answer.

Note that the above examples are simple because the coefficients of both the quadratic term and the constant term are prime numbers. This will not always be the case. If one or both of the terms are composite, the possible factors increase markedly. Be sure to always start by factoring the greatest common factor, if there is one. This will sometimes help make the trinomial simpler.

Example 2 Factor $6x^2 + 7x + 2$ completely.

Solution.

Always begin by first checking if there is a Greatest Common Factor (GCF). There is none. Next, look at the quadratic term. Notice that the quadratic term, $6x^2$ has two possible factors: $6x$ and x or $3x$ and $2x$. So the binomials we are looking for will have one of the following forms.

$$(6x\quad)(x\quad) \quad\text{or}\quad (3x\quad)(2x\quad)$$

Now look at the constant term, 2. Since it is prime and the linear term is positive, its only factors are 2 and 1. This means we have the following possible factors for the given trinomial.

a) $(6x+1)(x+2)$

b) $(6x+2)(x+1)$ - this is not a possibility because the binomial, $(6x+2)$, has a GCF of 2 and we already determined there was no GCF at the start.

c) $(3x+1)(2x+2)$ - this is not a possibility because the binomial, $(2x+2)$, has a GCF of 2 and we already determined there was no GCF at the start.

d) $(3x+2)(2x+1)$

This leaves us with only 2 possible factors. We check each of the two possibilities to see which is the correct factor.

$$
\begin{aligned}
(6x+1)(x+2) &= 6x(x+2)+1(x+2) \\
&= 6x^2+12x+x+2 \\
&= 6x^2+13x+2 \quad \times
\end{aligned}
$$

$$
\begin{aligned}
(3x+2)(2x+1) &= 3x(2x+1)+2(2x+1) \\
&= 6x^2+3x+4x+2 \\
&= 6x^2+7x+2 \quad \checkmark
\end{aligned}
$$

Therefore, $6x^2 + 7x + 2 = (3x+2)(2x+1)$.

We can summarize what happened in the last two examples as follows.

> If a trinomial does not have a common factor, then there can be no common factor in any of the binomials factors.

Example 3 Factor $6x^2 - 28x + 30$ completely.

Solution.
Always begin by first checking if there is a Greatest Common Factor (GCF). There is a GCF of 2. Factor 2 out to get: $6x^2 - 28x + 30 = 2(3x^2 - 14x + 15)$.

We will now factor $3x^2 - 14x + 15$. Look at the quadratic term. The quadratic term, $3x^2$ has only one possible factor: $3x$ and x. So the binomials we are looking for will have the following form.

$$(3x \quad)(x \quad)$$

Now look at the constant term. Since the constant term is positive while the linear term is negative, we will only consider negative factors of 15. The possible factors of 15 are: -1 and -15 or -3 and -5. This means we have the following possible factors for the given trinomial.

a) $(3x - 1)(x - 15)$

b) $(3x - 15)(x - 1)$ - this is not a possibility because the binomial, $(3x - 15)$, has a GCF of 3 and we already determined the GCF of 2 at the start.

c) $(3x - 3)(x - 5)$ - this is not a possibility because the binomial, $(3x - 3)$, has a GCF of 3 and we already determined the GCF of 2 at the start.

d) $(3x - 5)(x - 3)$

Check the two remaining possibilities to see which is the correct factor of $3x^2 - 14x + 15$.

$$\begin{aligned}
(3x - 1)(x - 15) &= 3x(x - 15) - 1(x - 15) \\
&= 3x^2 - 45x - x + 15 \\
&= 3x^2 - 46x + 15 \quad \times
\end{aligned}$$

$$\begin{aligned}
(3x - 5)(x - 3) &= 3x(x - 3) - 5(x - 3) \\
&= 3x^2 - 9x - 5x + 15 \\
&= 3x^2 - 14x + 15 \quad \checkmark
\end{aligned}$$

The completely factored form of the trinomial looks like this. Do not forget the GCF, 2.

$$\begin{aligned}
6x^2 - 28x + 30 &= 2(3x^2 - 14x + 15) \\
&= 2(3x - 5)(x - 3)
\end{aligned}$$

Verify that we have the correct factors.

$$2(3x-5)(x-3) = (6x-10)(x-3)$$
$$= 6x(x-3) - 10(x-3)$$
$$= 6x^2 - 18x - 10x + 30$$
$$= 6x^2 - 28x + 30 \checkmark$$

Exercise 3 Class Example

Factor the following trinomials completely. Be sure to check your answer.

a) $10x^2 + 9x + 2$

b) $12x^2 - 38x + 30$

Exercise 4 You Try

Factor the following trinomials completely. Be sure to check your answer.

a) $3x^2 + 11x + 6$

b) $12x^2 - 27x + 6$

Example 4 Factor $15x^2 + 29x + 8$ completely.

Solution.
Begin by first checking if there is a Greatest Common Factor (GCF). There is none.
Next, look at the quadratic term. The quadratic term, $15x^2$ has two possible factors: $15x$ and x or $5x$ and $3x$. So the binomials we are looking for will have one of the following form.

$$(15x \quad)(x \quad) \quad \text{or} \quad (5x \quad)(3x \quad)$$

Now look at the constant term. Since both the constant and linear term are positive, we will only consider positive factors of 8. The possible factors of 8 are: 1 and 8 or 2 and 4. This means we have the following possible factors (8 of them!!) for the given trinomial.

a) $(15x + 1)(x + 8)$ e) $(3x + 1)(5x + 8)$

b) $(15x + 8)(x + 1)$ f) $(3x + 8)(5x + 1)$

c) $(15x + 2)(x + 4)$ g) $(3x + 2)(5x + 4)$

d) $(15x + 4)(x + 2)$ h) $(3x + 4)(5x + 2)$

Since we found factors of the quadratic and constant term, the only one remaining to be verified is the linear term. To verify whether we have the correct linear term, we multiply the two outside terms and add the two inside terms of the binomial. These are the "O" and the "I" of "FOIL." Let us calculate the linear term of all the 8 possible factors to see which is the correct factor.

Possible Factors	Linear Term	
$(15x + 1)(x + 8)$	$(15x)(8) + (1)(x) = 121x$	$\times$
$(15x + 8)(x + 1)$	$(15x)(1) + (8)(x) = 23x$	$\times$
$(15x + 2)(x + 4)$	$(15x)(4) + (2)(x) = 62x$	$\times$
$(15x + 4)(x + 2)$	$(15x)(2) + (4)(x) = 34x$	$\times$
$(3x + 1)(5x + 8)$	$(3x)(8) + (1)(5x) = 29x$	$\checkmark$
$(3x + 8)(5x + 1)$	$(3x)(1) + (8)(5x) = 43x$	$\times$
$(3x + 2)(5x + 4)$	$(3x)(4) + (2)(5x) = 22x$	$\times$
$(3x + 4)(5x + 2)$	$(3x)(2) + (4)(5x) = 26x$	$\times$

Verify that we have the correct factors.

$$(3x + 1)(5x + 8) = 3x(5x + 8) + 1(5x + 8)$$
$$= 15x^2 + 24x + 5x + 8$$
$$= 15x^2 + 29x + 8 \quad \checkmark$$

Therefore, $15x^2 + 29x + 8 = (3x + 1)(5x + 8)$.

AC-Method

The number of cases to handle in the List and Check Method can get cumbersome. Some people prefer to use a different method. The method we describe below is based on reversing the steps of binomial multiplication. Let us multiply two binomials, working out each step of the distributive property.

$$(3x+4)(5x+2) = 3x(5x+2) + 4(5x+2) \qquad \text{Step (1)}$$
$$= 15x^2 + 6x + 20x + 8 \qquad \text{Step (2)}$$
$$= 15x^2 + 26x + 8 \qquad \text{Step (3)}$$

Notice the following.

- If we multiply the coefficient of the quadratic term and the coefficient of the linear term of our trinomial in Step (3), we get $15 \cdot 8 = 120$

- If we multiply the two coefficients of the linear terms in Step (2), we also get 120 ($6 \cdot 20 = 120$)

This observation will always be true and we can use these facts to reverse the steps. In order to do this, we must understand the following.

A) Rewrite the linear term of the trinomial as the sum of two appropriate terms (go from Step (3) to Step (2))

B) Factor the GCF from the first two terms and the GCF from the last two terms (go from Step (2) to Step (1))

C) Use $(5x+2)$ as a common factor (go from the left side of Step (1) to the right side of Step (1)).

Step B) and Step C) should be familiar. They are the instructions for *factoring by grouping* the 4-terms obtained in Step A).

Notice that a key part of this process is that we need to find a pair of numbers that multiply to 120 and add up to 26. In practice, this might take some time, since we might have to list all possible pairs of factors of 120 until we find a pair that adds up to 26.

This method for factoring a trinomial of the form $ax^2 + bx + c$ is known as the **ac-method**, since the first step will be to multiply the coefficient of the quadratic term, a, and the constant term, c, of the trinomial. Let us take a look at a few examples on how the *ac*-method is used to factor a trinomial.

Example 5 Factor $6x^2 - 19x + 15$ completely.

Solution.
Although the number of cases in this problem is still easy to handle using the list-and-check method, we will use the *ac*-method to illustrate the steps.
First see if there is a GCF. There is none. Let us factor the trinomial using the *ac*-method.

- Multiply $a = 6$ and $c = 15$ to get $6 \cdot 15 = 90$

- We need to find a pair of numbers that multiply to 90 and add up to -19. We find that the numbers, -9 and -10 fit the requirement. Use the pair of numbers to rewrite the linear term. That is, rewrite $-19x$ as $-9x + -10x$ or $-9x - 10x$.

- Use factoring by grouping to factor the trinomial.

$$
\begin{aligned}
6x^2 - 19x + 15 &= 6x^2 - 9x - 10x + 15 && \text{Rewrite linear term of trinomial} \\
&= 3x(2x - 3) - 5(2x - 3) && \text{Factor GCF of each group} \\
&= (2x - 3)(3x - 5) && \text{Our Factorization}
\end{aligned}
$$

Verify that we have the correct factors.

$$
\begin{aligned}
(2x - 3)(3x - 5) &= 2x(3x - 5) - 3(3x - 5) \\
&= 6x^2 - 10x - 9x + 15 \\
&= 6x^2 - 19x + 15 \ \checkmark
\end{aligned}
$$

Example 6 Factor $18x^2 + 45x + 25$ completely.

Solution.
First, see if there is a GCF. There is none. Next, we will use the ac-method to factor the trinomial.

- Multiply $a = 18$ and $c = 25$ to get $18 \cdot 25 = 450$

- We need to find a pair of numbers that multiply to 450 and add up to 45. Since the sum is positive 45, we will only consider positive factors of 450. Let us list the factors.

1 and 450	2 and 225	3 and 150	5 and 90	6 and 75
9 and 50	10 and 45	15 and 30	18 and 25	

- The numbers that multiply to 450 and add up to 45 are 15 and 30. Use the pair of numbers to rewrite the linear term. That is, rewrite $45x$ as $15x + 30x$.

- Use factoring by grouping to factor the trinomial.

$$
\begin{aligned}
18x^2 + 45x + 25 &= 18x^2 + 15x + 30x + 25 && \text{Rewrite linear term of trinomial} \\
&= 3x(6x + 5) + 5(6x + 5) && \text{Factor GCF of each group} \\
&= (6x + 5)(3x + 5) && \text{Our Factorization}
\end{aligned}
$$

Verify that we have the correct factors.

$$(6x+5)(3x+5) = 6x(3x+5)+5(3x+5)$$
$$= 18x^2+30x+15x+25$$
$$= 18x^2+45x+25 \checkmark$$

One important observation is that the order in which we write the two linear terms does not matter, because addition is commutative. To illustrate this, let us change the order of the linear terms that add up to 45x. That is, we will use $30x+15x$ instead of $15x+30x$.

$$18x^2+45x+25 = 18x^2+30x+15x+25 \qquad \text{Rewrite linear term of trinomial}$$
$$= 6x(3x+5)+5(3x+5) \qquad \text{Factor GCF of each group}$$
$$= (3x+5)(6x+5) \qquad \text{Our Factorization}$$

Verify that we have the correct factors.

$$(3x+5)(6x+5) = 3x(6x+5)+5(6x+5)$$
$$= 18x^2+15x+30x+25$$
$$= 18x^2+45x+25$$

Since multiplication is commutative, $18x^2+45x+25 = (6x+5)(3x+5) = (3x+5)(6x+5)$

Example 7 Factor $6x^2+19x-36$ completely.

Solution.
First, see if there is a GCF. There is none. We will use the *ac*-method to factor the trinomial.

- Multiply $a=6$ and $c=-36$ to get $18 \cdot -36 = -216$

- We need to find a pair of numbers that multiply to -216 and add up to 19. In this case, we will find a pair of numbers that multiply to 216. We will take care of the negative sign later. Let us list the factors of 216.

1 and 216	2 and 108	3 and 72	4 and 54
6 and 36	8 and 27	9 and 24	12 and 18

- Since the product of *ac* is negative, the pair of numbers we are looking for must have opposite signs. That means that we are looking for a pair of numbers that multiply to -216 with a **difference** of 19. The pair -8 and 27 fit this requirement. Rewrite 19x as $-8x+27x$.

- Use factoring by grouping to factor the trinomial.

$$6x^2 + 19x - 36 = 6x^2 - 8x + 27x - 36 \qquad \text{Rewrite linear term of trinomial}$$
$$= 2x(3x - 4) + 9(3x - 4) \qquad \text{Factor GCF of each group}$$
$$= (3x - 4)(2x + 9) \qquad \text{Our Factorization}$$

One useful application of the *ac*-method is that it gives us a tool to determine if a trinomial is prime.

Example 8 Factor $18x^2 + 75x + 90$ completely.

Solution.
First see if there is a GCF. In this case, there is a GCF of 3. Factor out the 3, to get, $18x^2 + 75x + 90 = 3(6x^2 + 25x + 10)$.
Next, we will use the *ac*-method to factor the trinomial $6x^2 + 25x + 10$.

- Multiply $a = 6$ and $c = 30$ to get $6 \cdot 30 = 180$

- We need to find a pair of numbers that multiply to 180 and add up to 25. Since the sum is positive, we will only consider positve factors of 180. Let us list the factors.

1 and 180	2 and 90	3 and 60	4 and 45	5 and 36
6 and 30	9 and 20	10 and 18	12 and 15	

We see that none of the pairs add up to 25. We conclude that the trinomial $6x^2 + 25x + 30$ is prime. Therefore, $18x^2 + 75x + 90 = 3(6x^2 + 25x + 30)$.

Exercise 5 **Class Example**
Factor the following trinomials completely.

a) $12x^2 + 17x + 6$ \qquad\qquad\qquad b) $8x^2 - 16x - 10$

Exercise 6 **Class Example**

Factor the following trinomials completely.

a) $8x^2 - 2x - 15$

b) $6x^2 - x + 30$

6.3: Exercises

Factor the trinomial if possible. If not, state why it is prime.

1. $4x^2 + 4x - 3$

2. $5x^2 + 32x + 12$

3. $3x^2 + 4x - 8$

4. $6x^2 + 7x - 3$

5. $7x^2 - 11x - 6$

6. $9x^2 - 6x - 8$

7. $8x^2 + 2x - 3$

8. $10x^2 + 9x + 2$

9. $15x^2 + 14x + 3$

10. $2x^2 + 25x + 12$

11. $12x^2 - 2x - 4$

12. $21x^2 - 5x - 6$

13. $28x^2 - 13x - 6$

14. $8x^2 + 14x - 15$

15. $20x^2 + 7x - 6$

6.4   Factoring Special Products

Objective: To factor the difference of two squares and recognize perfect square trinomials

Difference of Two Squares

We will learn how to recognize the pattern for the difference of two squares, $a^2 - b^2$, so that we can quickly factor these expressions. Let us take a look at an example using the strategies learned in the previous sections.

Example 1 Factor $x^2 - 16$ completely.

Solution.
You may notice that this is a binomial rather than a trinomial. It is missing a linear term. We will *pencil* in the linear term with a coefficient of 0, giving us $x^2 + \mathbf{0x} - 16$. We need to find factors of -16 that sum to 0. Since the constant term is negative, we will only consider a positive and a negative factor.

Factors of -16	Product of Factors	Sum of Factors
$-1, 16$	$(-1) \cdot (16) = -16$	$(-1) + (16) = 15$
$1, -16$	$(1) \cdot (-16) = -16$	$(1) + (-16) = -15$
$-4, 4$	$(-4) \cdot (4) = -16$	$(-4) + (4) = 0$
$4, -4$	$(4) \cdot (-4) = -16$	$(4) + (-4) = 0$

The last two pairs give us what we need. We put those values into the binomials to get our final answer.

$$x^2 - 16 = (x - 4)(x + 4)$$
$$x^2 - 16 = (x + 4)(x - 4)$$

Either answer is correct because multiplication is commutative. To check our answer, we perform binomial multiplication and confirm we get back the original trinomial.

$$
\begin{aligned}
(x - 4)(x + 4) &= x(x + 4) - 4(x + 4) \\
&= x^2 + 4x - 4x - 16 \\
&= x^2 - 16 \ \checkmark
\end{aligned}
$$

$$
\begin{aligned}
(x + 4)(x - 4) &= x(x - 4) + 4(x - 4) \\
&= x^2 - 4x + 4x - 16 \\
&= x^2 - 16 \ \checkmark
\end{aligned}
$$

Let us see take a look at a more general case but first, let us review the product of the following factors.

$$(a+b)(a-b) = a(a-b) + b(a-b)$$
$$= a^2 - ab + ab - b^2$$
$$= a^2 - b^2$$

Notice that the middle terms are opposites of each other and when combined, will equal 0, leaving us with the difference of two perfect squares. Recognizing this pattern will help us easily factor the difference of two squares.

Difference of Two Squares

$$a^2 - b^2 = (a+b)(a-b)$$

Example 2 Factor $m^2 - 64$ completely.

Solution.

$$m^2 - 64 = (m)^2 - (8)^2 \qquad \text{Rewrite binomial as a difference of two squares}$$
$$= (m+8)(m-8) \qquad \text{Our Factorization}$$

Check that we have the correct factors.

$$(m+8)(m-8) = m(m-8) + 8(m-8)$$
$$= m^2 - 8m + 8m - 64$$
$$= m^2 - 64 \ \checkmark$$

Warning. It is important to note that the Sum of Two Squares, $a^2 + b^2$ is not factorable.

Sum of Two Squares:

$$a^2 + b^2 \text{ is Prime.}$$

Let us take a look at an example to see why the sum of two squares is not factorable.

Example 3 Factor $x^2 + 16$ completely. If the binomial is not factorable, identify it as prime.

Solution.
Notice the linear term is missing. We will *pencil* in the linear term with a coefficient of 0.

Consider then $x^2 + 0x + 16$. We need to find factors of 16 that sum to 0. Since the constant term is positive and the sum is 0, we will consider only positive factors or only negative factors.

Factors of 16	Product of Factors	Sum of Factors
$1, 16$	$1 \cdot 16 = 16$	$1 + 16 = 17$
$-1, -16$	$(-1) \cdot (-16) = 16$	$(-1) + (-16) = -17$
$4, 4$	$4 \cdot 4 = 16$	$4 + 4 = 8$
$-4, -4$	$(-4) \cdot (-4) = 16$	$(-4) + (-4) = -8$

We have exhausted all the possible factors of 16 and none of them add up to 0. $x^2 + 16$ is not factorable and is therefore, prime.

What if the binomial is written in ascending order instead of descending order? If the binomial fits the characteristics of a difference of squares, we follow the above factoring strategy. There is no need to rearrange the order.

Example 4 Factor $9 - n^2$ completely.

Solution.

$$9 - n^2 = (3)^2 - (n)^2 \qquad \text{Rewrite binomial as a difference of two squares}$$
$$= (3 + n)(3 - n) \qquad \text{Our Factorization}$$

Check that we have the correct factors.

$$(3 + n)(3 - n) = 3(3 - n) + n(3 - n)$$
$$= 9 - 3n + 3n - n^2$$
$$= 9 - n^2 \ \checkmark$$

Exercise 1 **Class Example**
Factor the following binomials completely, if possible. If the binomial is not factorable, identify it as prime. Be sure to check your answer.

a) $p^2 - 4$

b) $y^2 + 49$

Exercise 2 Class Example

Factor the following binomials completely, if possible. If the binomial is not factorable, identify it as prime. Be sure to check your answer.

a) $36 - n^2$

c) $81 + a^2$

b) $3m^2 - 27$

d) $x^4 - 16$

Exercise 3 You Try

Factor the following binomials completely, if possible. If the binomial is not factorable, identify it as prime. Be sure to check your answer.

a) $25p^2 - 9$

c) $h^2 + 100$

b) $1 - y^2$

d) $2c^2 - 32$

Exercise 4 **You Try**

Factor the following binomials completely, if possible. If the binomial is not factorable, identify it as prime. Be sure to check your answer.

a) $36 - 4w^2$

b) $x^4 - 81$

Perfect Square Trinomial

In previous sections, we practiced factoring trinomials of the form, $ax^2 + bx + c$. A trinomial such as $x^2 + 10x + 25$, has a special pattern. The factors of this expression are two identical binomials, $(x+5)(x+5)$. We can rewrite these factors using the exponential notation, $(x+5)^2$. Notice that it is a perfect square.

It is not always easy to see this pattern. We can always use our previous methods for factoring trinomials. However, if we are able to recognize the perfect square pattern, we can eliminate some steps and save time.

Let us recall from our work on multiplication of binomials.

$$(a+b)(a+b) = a^2 + 2ab + b^2 \text{ and } (a-b)(a-b) = a^2 - 2ab + b^2$$

To recognize this pattern, first determine if the sign pattern of the binomial will be the same. The pattern in the original trinomial must be either the following.

$$(\_+\_+\_) \quad \text{or} \quad (\_-\_-\_)$$

Next, notice that the quadratic term and the constant term are perfect squares. If the quadratic term and the constant term are perfect squares, and twice the product of the square root of each of these terms gives the linear term of the trinomial, then you have a pattern for a perfect square trinomial.

Perfect Square Trinomial:

$$a^2 + 2ab + b^2 = (a+b)^2 \quad \text{or} \quad a^2 - 2ab + b^2 = (a-b)^2$$

Example 5 Factor $x^2 - 6x + 9$ completely.

Solution.

Notice that the quadratic term, x^2, and the constant term, 9, are perfect squares. We can rewrite $x^2 = (x)^2$. Since the given linear term is negative, we rewrite $9 = (-3)^2$. When x and -3 are multiplied together, we get the product $-3x$. When $-3x$ is multiplied by 2, we get the linear term, $-6x$, which matches the linear term of the given trinomial. This indicates that the given trinomial is a perfect square trinomial.

$$
\begin{aligned}
x^2 - 6x + 9 &= (x)^2 - 2(x)(3) + (-3)^2 \\
&= (x-3)(x-3) \\
&= (x-3)^2
\end{aligned}
$$

Verify that we have the correct factors.

$$
\begin{aligned}
(x-3)^2 &= (x-3)(x-3) \\
&= x(x-3) - 3(x-3) \\
&= x^2 - 3x - 3x + 9 \\
&= x^2 - 6x + 9 \ \checkmark
\end{aligned}
$$

Example 6 Factor $25x^2 + 10x + 1$ completely.

Solution.
Notice that the quadratic term, $25x^2$, and the constant term, 1, are perfect squares. We can rewrite $25x^2 = (5x)^2$. Since the given linear term is positive, we rewrite $1 = (1)^2$. When $5x$ and 1 are multiplied together, we get the product $5x$. When $5x$ is multiplied by 2, we get the linear term, $10x$, which matches the linear term of the given trinomial. This indicates that the given trinomial is a perfect square trinomial.

$$
\begin{aligned}
25x^2 + 10x + 1 &= (5x)^2 + 2(5x)(1) + (1)^2 \\
&= (5x+1)(5x+1) \\
&= (5x+1)^2
\end{aligned}
$$

Verify that we have the correct factors.

$$
\begin{aligned}
(5x+1)^2 &= (5x+1)(5x+1) \\
&= 5x(5x+1) + 1(5x+1) \\
&= 25x^2 + 5x + 5x + 1 \\
&= 25x^2 + 10x + 1 \ \checkmark
\end{aligned}
$$

Exercise 5 **Class Example**
Factor the following trinomials completely, if possible. Be sure to check your answer.

a) $m^2 - 12m + 36$ b) $4y^2 + 20y + 25$

Exercise 6 You Try

Factor the following binomials completely, if possible. Be sure to check your answer.

a) $y^2 + 8y + 16$ 　　　　　　　　　　　　　c) $49p^2 - 28p + 4$

b) $9x^2 + 12x + 4$ 　　　　　　　　　　　　　d) $25 - 30k + 9k^2$

Example 7 Factor $16m^3 - 24m^2 + 9m$ completely.

Solution.

Always start factoring by looking for the GCF. The GCF in this case is m.

$$16m^3 - 24m^2 + 9m = m(16m^2 - 24m + 9)$$

Next, work on the trinomial, $16m^2 - 24m + 9$. Notice that the quadratic term, $16m^2$, and the constant term, 9, are perfect squares. We can rewrite $16m^2 = (4m)^2$. Since the given linear term is negative, we rewrite $9 = (-3)^2$. When 4m and -3 are multiplied together, we get the product $-12m$. When $-12m$ is multiplied by 2, we get the linear term, $-24m$, which matches the linear term of the given trinomial. This indicates that the given trinomial is a perfect square trinomial.

$$\begin{aligned}
16m^3 - 24m^2 + 9m &= m(16m^2 - 24m + 9) \\
&= m[(4m)^2 - 2(4m)(3) + (-3)^2] \\
&= m[(4m - 3)(4m - 3)] \\
&= m(4m - 3)^2
\end{aligned}$$

Check that we have the correct factors.

$$m(4m-3)^2 = m(4m-3)(4m-3)$$
$$= (4m^2-3m)(4m-3)$$
$$= 4m^2(4m-3)-3m(4m-3)]$$
$$= 16m^3-12m^2-12m^2+9m]$$
$$= 16m^3-24m^2+9m \quad \checkmark$$

Exercise 7 Class Example

Factor the following trinomials completely. Be sure to check your answer.

a) $50n^2-40n+8$ b) $4m^3+20m^2+25m$

Exercise 8 You Try

Factor the following trinomials completely, if possible. Be sure to check your answer.

a) $5y^2-40y+80$ b) $36x^2+24x+4$

The following summarizes factoring special products.

- Difference of Two Squares: $a^2-b^2 = (a+b)(a-b)$

- Sum of Two Squares: a^2+b^2 is Prime

- Perfect Square Trinomial: $a^2+2ab+b^2 = (a+b)^2$

- Perfect Square Trinomial: $a^2-2ab+b^2 = (a-b)^2$

As always, when factoring, it is important to first check for the GCF. Only after checking for the GCF should we be using the special products.

6.4: Exercises

Factor each trinomial completely, if possible. If the trinomial is not factorable, identify it as prime.

1. $x^2 - 9$

2. $x^2 - 1$

3. $v^2 + 25$

4. $4 - p^2$

5. $4v^2 - 1$

6. $9k^2 - 4$

7. $1 - 9a^2$

8. $3x^2 - 27$

9. $125x^2 + 45$

10. $5n^2 - 20$

11. $18a^2 - 50$

12. $64 + 4m^2$

13. $a^2 - 2a + 1$

14. $k^2 + 4k + 4$

15. $n^2 - 8n + 16$

16. $25p^2 - 10p + 1$

17. $4k^2 + 28k + 49$

18. $x^2 + 8x + 16$

19. $25a^2 + 30a + 9$

20. $4a^2 - 20a + 25$

21. $18m^2 - 24m + 8$

22. $5x^2 + 10x + 5$

23. $20x^2 + 20x + 5$

24. $8x^2 - 24x + 18$

25. $a^4 - 81$

26. $n^4 - 1$

27. $16 - z^4$

Rescue Roody!

28. Roody was told to factor $16x^2 - 36$ completely. Roody recognized that the binomial is a difference of two squares. This is how Roody factored the binomial.

$$16x^2 - 36 = (4x + 6)(4x - 6)$$

Roody then checked his work.

$$(4x + 6)(4x - 6) = 4x(4x - 6) + 6(4x - 6)$$
$$= 16x^2 - 24x + 24x - 36$$
$$= 16x^2 - 36$$

Everything seems fine, but his answer was marked incorrect. Help Roody.

6.5 Factoring Strategies

Objective: To be able to use factoring strategies to factor various polynomials

With so many different methods to factor, it is easy to get lost as to which method to use and when. In this section we will organize the different factoring strategies we have seen.

Factoring Strategies

1. Always factor the Greatest Common Factor (GCF) first.

2. Look at the number of terms in the polynomial.

    (a) Binomial

        i. A Sum of Squares: $a^2 + b^2$ is prime

        ii. A Difference of Squares: $a^2 - b^2 = (a+b)(a-b)$

    (b) Trinomial

        i. Use the strategies discussed in the previous sections to factor trinomials of the form $ax^2 + bx + c$

        ii. Perfect Square Trinomial
            • $a^2 + 2ab + b^2 = (a+b)^2$
            • $a^2 - 2ab + b^2 = (a-b)^2$

    (c) 4-term Polynomial

        i. Factor by Grouping

It is important to be comfortable and confident using the factoring methods and deciding which method to use.

Example 1 Factor $7ax - 14x + 3a - 6$ completely.

Solution.
The polynomial does not have a GCF. Since it is a 4-term polynomial, we will factor by grouping.

$$7ax - 14x + 3a - 6 = \underbrace{7ax - 14x} + \underbrace{3a - 6} \qquad \text{Group first 2 terms and last 2 terms}$$
$$= 7x(a-2) + 3(a-2) \qquad \text{Find GCF for each group and factor}$$
$$= (a-2)(7x+3) \qquad \text{Our Factorization}$$

Verify that we have the correct factors.

$$(a-2)(7x+3) = a(7x+3) - 2(7x+3)$$
$$= 7ax + 3a - 14x - 6$$
$$= 7ax - 14x + 3a - 6 \quad \checkmark$$

Example 2 Factor $a^2 - 22a - 48$ completely.

Solution.
The trinomial has no GCF. We need to find factors of -48 that sum up to -22. Given that the product is negative, we have to consider combinations of positive and negative factors.

Factors of -48	Product of Factors	Sum of Factors
$1, -48$	$1 \cdot (-48) = -48$	$1 + (-48) = -47$
$2, -24$	$2 \cdot (-24) = -48$	$2 + (-24) = -22$

We can stop at this point since we have found the pair of numbers we need. The trinomial factors as follows.

$$a^2 - 22a - 48 = (a+2)(a-24)$$

Check that we have the correct factors.

$$\begin{aligned} (a+2)(a-24) &= a(a-24) + 2(a-24) \\ &= a^2 - 24a + 2a - 48 \\ &= a^2 - 22a - 48 \ \checkmark \end{aligned}$$

Example 3 Factor $100y^2 - 400$ completely.

Solution.
The binomial has a GCF of 100. Factor 100 out and then factor the resulting binomial, which is a difference of two squares.

$$\begin{aligned} 100y^2 - 400 &= 100(y^2 - 4) \\ &= 100(y+2)(y-2) \qquad \text{Our Factorization} \end{aligned}$$

Check to verify factors are correct.

$$\begin{aligned} 100(y+2)(y-2) &= (100y + 200)(y-2) \\ &= 100y(y-2) + 200(y-2) \\ &= 100y^2 - 200y + 200y - 400 \\ &= 100y^2 - 400 \ \checkmark \end{aligned}$$

Example 4 Factor $6n^3 + 14n^2 - 2n$ completely.

Solution.
Begin by first checking if there is a Greatest Common Factor (GCF). We see that $2n$ is a GCF. Factor it out to get $6n^3 + 14n^2 - 2n = 2n(3n^2 + 7n - 1)$.

Next, look at the quadratic term of the trinomial in the parenthesis. The quadratic term, $3n^2$

has one possible factor: $3n$ and n. So the binomials we are looking for will have the following form.

$$(3n \quad)(n \quad)$$

Now look at the constant term, -1. Since the constant term is negative, we will consider positive and negative factors of -1. The possible factors of -1 are: 1 and -1. This means we have the following possible factors for the trinomial, $3n^2 + 7n - 1$.

a) $(3n+1)(n-1)$ b) $(3n-1)(n+1)$

Since we found factors of the quadratic and constant term, the only one remaining to be verified is the linear term, $7n$. To verify whether we have the correct linear term, we multiply the two outside terms and add the two inside terms of the binomial. These are the "O" and the "I" of "FOIL." Let us calculate the linear term of all the possible factors to see which is the correct factor.

Possible Factors	Linear Term	
$(3n+1)(n-1)$	$(3n)(-1)+(1)(n) = -2n$	$\times$
$(3n-1)(n+1)$	$(3n)(1)+(-1)(n) = 2n$	$\times$

We have exhausted all the possible factors but none of them gave the linear term we are looking for. The trinomial, $3n^2 + 7n - 1$, must be prime.

Therefore, $6n^3 + 14n^2 - 2n = 2n(3n^2 + 7n - 1)$.

Exercise 1 Class Example
Factor the following expressions completely. If the expression is not factorable, identify it as prime.

a) $4x^2 + 56x + 196$ b) $ab + 3a - 7b - 21$

Exercise 2 Class Example

Factor the following expressions completely. If the expression is not factorable, identify it as prime.

a) $5 - 45m^3$

b) $36p^4 - 13p^2 + 1$

Exercise 3 You Try

Factor the following expressions completely. If the expression is not factorable, identify it as prime.

a) $3m^2 + 3m - 18$

c) $15p^2 + p - 2$

b) $4n^2 - 8n - 4$

d) $2x^3 - x^2 - 2x + 1$

6.5: Exercises

Factor each expression completely. If not factorable, identify as prime.

1. $2p(3p+4)+(3p+4)$

2. x^2+4x+3

3. m^2-4

4. v^3+v

5. n^2-3n+4

6. $2x^2-11x+15$

7. $2x^2-10x+12$

8. $ab-3a+2b-6$

9. n^3-5n^2-6n

10. $x^2+8x+16$

11. $5x^2-22x-15$

12. n^3+7n^2+10n

13. $8y^3+6y^2+20y+15$

14. $16a^2-9$

15. $5n^2+7n-6$

16. $16x^2-72x+81$

17. $2k^2+k-10$

18. $9n^3-6n^2+3n$

19. $2x^3+6x^2-20x$

20. $3xy-15y-x+5$

21. $9-25y^2$

22. $4x^2+24xy+36y^2$

23. $3k^3-27k^2+60k$

24. $5u^2-9u+4$

25. $45m^2-150m+125$

26. $27m^2-48$

27. x^4+4x^2

28. $n-n^3$

29. p^4-81

30. $y^3+5y^2-4y-20$

6.6 Solving Quadratic Equations by Factoring

Objective: To be able to use factoring techniques to solve quadratic equations

We begin this section by stating a basic fact about real numbers: *if the product of two numbers is zero, then at least one of the factors must be equal to zero.* This is known as the zero-product property, and can be written as follows.

Zero-Product Property

If $a \cdot b = 0$

then either $a = 0$ or $b = 0$

Example 1 Use the zero-product property to solve the equation $(x+1)(x-2)=0$. Check your answers.

Solution.
The zero-product property tells us that if $(x+1)(x-2) = 0$, then we must have

$$x + 1 = 0 \text{ or } x - 2 = 0$$

We now have two linear equations to solve. Each equation will give us a solution to our original problem.

First Solution
$$\begin{aligned} x + 1 &= 0 \qquad &&\text{Subtract 1 from each side}\\ x &= -1 &&\text{Our Solution} \end{aligned}$$
Second Solution
$$\begin{aligned} x - 2 &= 0 \qquad &&\text{Add 2 to each side}\\ x &= 2 &&\text{Our Solution} \end{aligned}$$

Verify our solutions by substituting the values we got into the original equation.

Check $x = -1$:

$$(-1+1)(-1-2) \overset{?}{=} 0$$
$$(0)(-3) = 0 \ \checkmark$$

Check $x = 2$:

$$(2+1)(2-2) \overset{?}{=} 0$$
$$(3)(0) = 0 \ \checkmark$$

The solutions to the equation are $x = -1$ and $x = 2$.

On our previous example the original equation was given in factored form. Usually this is not be the case. To be able to solve quadratic equations and use the zero-product property, the equation has to be set equal to zero and the quadratic expression needs to be in factored form.

Example 2 Solve the equation $2x^2 + 6x = 0$. Check your answers.

Solution.
The equation is set to 0. Next, factor the quadratic expression.

$$2x^2 + 6x = 0 \qquad\qquad \text{Factor GCF}$$
$$2x(x+3) = 0 \qquad\qquad \text{Apply zero-product property}$$

By the zero-product property, we now have two linear equations to solve. Each equation will give us a solution to our original problem.

First Solution
$$2x = 0 \qquad\qquad \text{Divide each side by 2}$$
$$x = 0 \qquad\qquad \text{Our Solution}$$

Second Solution
$$x + 3 = 0 \qquad\qquad \text{Subtract 3 from each side}$$
$$x = -3 \qquad\qquad \text{Our Solution}$$

Verify our solutions by substituting the values we got into the original equation, $2x^2 + 6x = 0$.

Check $x = 0$:

$$2(0)^2 + 6(0) \overset{?}{=} 0$$
$$2(0) + 6(0) \overset{?}{=} 0$$
$$0 + 0 = 0 \ \checkmark$$

Check $x = -3$:

$$2(-3)^2 + 6(-3) \overset{?}{=} 0$$
$$2(9) + 6(-3) \overset{?}{=} 0$$
$$18 + -18 = 0 \ \checkmark$$

The solutions to the quadratic equations are $x = 0$ and $x = -3$.

Exercise 1 Class Example
Solve the following equations. Be sure to check your solutions.

a) $(x-1)(2x+1) = 0$

b) $3x^2 - 15x = 0$

Exercise 2 **You Try**
Solve the following equations. Be sure to check your solutions.

a) $(3x - 4)(x + 5) = 0$ b) $2x^2 - 8x = 0$

Example 3 Solve the equation $x^2 + 5x - 6 = 0$. Check your answers.

Solution.
The equation is set to 0. Next, factor the quadratic expression.

$$x^2 + 5x - 6 = 0 \qquad \text{Factor the trinomial}$$
$$(x + 6)(x - 1) = 0 \qquad \text{Apply zero-product property}$$

By the zero-product property, we now have two linear equations to solve. Each equation will give us a solution to our original problem.

First Solution
$$x + 6 = 0 \qquad \text{Subtract 6 from each side}$$
$$x = -6 \qquad \text{Our Solution}$$

Second Solution
$$x - 1 = 0 \qquad \text{Add 1 to each side}$$
$$x = 1 \qquad \text{Our Solution}$$

Verify that we have the correct solutions by substituting the values we got into the original equation, $x^2 + 5x - 6 = 0$.

Check $x = -6$:   Check $x = 1$:

$$(-6)^2 + 5(-6) - 6 \overset{?}{=} 0$$
$$36 - 30 - 6 \overset{?}{=} 0$$
$$6 - 6 = 0 \ \checkmark$$

$$(1)^2 + 5(1) - 6 \overset{?}{=} 0$$
$$1 + 5 - 6 \overset{?}{=} 0$$
$$6 - 6 = 0 \ \checkmark$$

The solutions to the quadratic equations are $x = -6$ and $x = 1$.

Exercise 3 Class Example
Solve the equation $x^2 - 5x - 14 = 0$. Be sure to check your solutions.

Exercise 4 You Try
Solve the equation $x^2 + x - 12 = 0$. Be sure to check your solutions.

Notice that to use the zero-product property, we need to have a zero on one side of the equation and the polynomial expression on the other side. If that is not the case, we must first rewrite the equation to have it in this form.

Example 4 Solve the equation $x^2 + 7x = 18$. Check your answers.

Solution.
First, set the equation to 0. In this case, subtract 18 from each side of the equation to get $x^2 + 7x - 18 = 0$. Next, factor the quadratic expression.

$$x^2 + 7x - 18 = 0 \qquad \text{Factor the trinomial}$$
$$(x + 9)(x - 2) = 0 \qquad \text{Apply zero-product property}$$

By the zero-product property, we now have two linear equations to solve. Each equation will give us a solution to our original problem.

First Solution

$$x + 9 = 0 \qquad \text{Subtract 9 from each side}$$
$$x = -9 \qquad \text{Our Solution}$$

Second Solution

$$x - 2 = 0 \qquad \text{Add 2 to each side}$$
$$x = 2 \qquad \text{Our Solution}$$

Verify our solutions by substituting the values we got into the original equation, $x^2 + 7x = 18$.

Check $x = -9$:

$$(-9)^2 + 7(-9) \stackrel{?}{=} 18$$
$$81 + 7(-9) \stackrel{?}{=} 18$$
$$81 - 63 = 18 \ \checkmark$$

Check $x = 2$:

$$(2)^2 + 7(2) \stackrel{?}{=} 18$$
$$4 + 7(2) \stackrel{?}{=} 18$$
$$4 + 14 = 18 \ \checkmark$$

The solutions to the quadratic equations are $x = -9$ and $x = 2$.

Exercise 5 Class Example
Solve the following equations. Be sure to check your solutions.

a) $m^2 + 8m = 20$

b) $2x^2 + 3x = 12 - 2x$

Exercise 6 You Try
Solve the following equations.Be sure to check your solutions.

a) $x^2 + 4x = 21$

b) $3y^2 = 6y$

Example 5 Solve the equation $x^2 = 25$. Check your answers.

Solution.

First, set the equation to 0. In this case, subtract 25 from each side of the equation to get $x^2 - 25 = 0$. Next, factor the quadratic expression.

$$x^2 - 25 = 0 \qquad \text{Factor the quadratic expression}$$
$$(x+5)(x-5) = 0 \qquad \text{Apply zero-product property}$$

By the zero-product property, we now have two linear equations to solve. Each equation will give us a solution to our original problem.

First Solution
$$x + 5 = 0 \qquad \text{Subtract 5 from each side}$$
$$x = -5 \qquad \text{Our Solution}$$

Second Solution
$$x - 5 = 0 \qquad \text{Add 5 to each side}$$
$$x = 5 \qquad \text{Our Solution}$$

Verify our solutions by substituting the values we got into the original equation, $x^2 = 25$.

Check $x = -5$:   Check $x = 5$:

$$(-5)^2 \stackrel{?}{=} 25$$
$$25 = 25 \ \checkmark$$

$$(5)^2 \stackrel{?}{=} 25$$
$$25 = 25 \ \checkmark$$

The solutions to the quadratic equations are $x = -5$ and $x = 5$.

Example 6 Solve the equation $x(x+1) = 12$. Check your answers.

Solution.

Remember that you cannot use the zero-product property unless the equation is set to zero and the polynomial is factored.

$$x(x+1) = 12 \qquad \text{Subtract 12 from each side}$$
$$x(x+1) - 12 = 0 \qquad \text{Apply order of operation by distributing } x$$
$$x^2 + x - 12 = 0 \qquad \text{Factor Trinomial}$$
$$(x+4)(x-3) = 0 \qquad \text{Apply zero-product property}$$

By the zero-product property, we now have two linear equations to solve. Each equation will give us a solution to our original problem.

First Solution

$$x + 4 = 0 \qquad \text{Subtract 4 from each side}$$
$$x = -4 \qquad \text{Our Solution}$$

Second Solution

$$x - 3 = 0 \qquad \text{Add 3 to each side}$$
$$x = 3 \qquad \text{Our Solution}$$

Verify our solutions by substituting the values we got into the original equation, $x(x+1) = 12$.

Check $x = -4$:   Check $x = 3$:

$$(-4)(-4+1) \overset{?}{=} 12$$
$$(-4)(-3) = 12 \ \checkmark$$

$$(3)(3+1) \overset{?}{=} 12$$
$$(3)(4) = 12 \ \checkmark$$

The solutions to the quadratic equations are $x = -4$ and $x = 3$.

Exercise 7 Class Example
Solve the following equations. Be sure to check your solutions.

a) $x^2 = 49$

c) $x(2x+5) = 3$

b) $x(x+3) = 4$

d) $(x+2)^2 + x^2 = 10$

Exercise 8 You Try

Solve the following equations. Be sure to check your solutions.

a) $2x^2 = 8$

c) $x(3x - 5) = 2$

b) $x(x + 6) = 7$

d) $(x + 1)(x - 2) = 4$

Let us now take a look at how we can find a quadratic equation given its solution.

Example 7 Find a quadratic equation with integer coefficients that has solutions $x = 3$ and $x = -5$.

Solution.

We can assume that the quadratic equation we are looking for has the form $ax^2 + bx + c = 0$. If we want $x = 3$ to be one of the solutions to our equation, then $(x - 3)$ must be a factor of our quadratic expression. Similarly, if we want $x = -5$ to be a solution to our equation, then $(x + 5)$ must be a factor of our quadratic expression.

Therefore, the equation we want must look like:

$$(x - 3)(x + 5) = 0 \qquad \text{Factors of the quadratic expression}$$
$$x^2 + 5x - 3x - 15 = 0 \qquad \text{Apply distributive property}$$
$$x^2 + 2x - 15 = 0 \qquad \text{Combine like terms}$$

Exercise 9 **Class Example**

Find a quadratic equation with integer coefficients that has solutions $x = 4$ and $x = -\dfrac{3}{2}$

Exercise 10 **You Try**

Find a quadratic equation with integer coefficients that has solutions $x = -2$ and $x = \dfrac{3}{4}$

6.6: Exercises

Solve each equation by factoring.

1. $x^2 + 5x + 6 = 0$

2. $x^2 - 6x = 0$

3. $p^2 + 3p = 0$

4. $t^2 = t$

5. $5x^2 + 6x = 0$

6. $6m^2 = 12m$

7. $x^2 + 4x + 3 = 0$

8. $m^2 - 2m = 8$

9. $p^2 - 10p + 24 = 0$

10. $n^2 - 6n = 0$

11. $2x^2 - 14x + 24 = 0$

12. $2x^2 = x + 6$

13. $a^2 + a = 6$

14. $n^2 - 15n + 36 = 0$

15. $x^2 - 23x + 130 = 0$

16. $5y^2 - 75y = -220$

17. $n^2 + 13n = -30$

18. $p^3 - p = 0$

19. $4x^2 + 4x + 1 = 0$

20. $n^2 + 7n + 6 = 0$

21. $2x^2 = 11x - 5$

22. $6k^2 - 17k = -5$

23. $9n^2 + 6n = -1$

24. $x^3 + 6x^2 = 0$

25. $x^2 - 15x = 250$

Give a quadratic equation with integer coefficients that has the given solutions.

26. $x = 5$ and $x = 1$

27. $x = 3$ and $x = -5$

28. $x = -4$ and $x = 4$

29. $x = 7$ and $x = -3$

30. $x = 3$

31. $x = 4$

32. $x = \frac{2}{3}$ and $x = 4$

33. $x = \frac{3}{2}$ and $x = -\frac{1}{3}$

34. $x = 5$ and $x = \frac{1}{10}$

35. $x = -\frac{4}{5}$

Chapter 6 Assessment

Factor each expression completely.

1. $6x^3 + 14x^2$

2. $7c^3 - 14c^2 + 3c - 6$

3. $y^2 + 7y + 6$

4. $x^2 - 5x - 24$

5. $n^2 + 49$

6. $2y^2 + y - 3$

7. $4p^2 - 4p + 1$

8. $3c^2 + 24c + 48$

9. $1 - 25m^2$

10. $5a^3 - 5a^2 - 3a + 3$

Solve each of the following equations by factoring.

11. $(3k + 4)(k - 5) = 0$

12. $x^2 + 12x + 20 = 0$

13. $p^2 + 4p = 0$

14. $m^2 + 9 = 6m$

15. $y(3y - 1) = 10$

16. $3w^2 = 10w - 3$

17. $2y^2 = 8$

18. $4n^2 = 12n$

7. Radicals Part I

7.1 Square Roots

Objective: To understand the meaning of square root and to differentiate between rational versus irrational numbers, exact and approximate values

You may already know that the square root of 4 is 2 or that $\sqrt{25} = 5$. Let us formalize our understanding of square root.

$$\sqrt{b} = a \quad \text{for any} \quad b > 0 \quad \text{if} \quad a^2 = b$$

In other words, $\sqrt{25} = 5$ because $5^2 = 25$.

Some important things to note:

- The symbol $\sqrt{}$ is known as the **radical**.

- The expression under the radical symbol is known as the **radicand**. For example, b is the radicand of $\sqrt{b}$.

- $\sqrt{b}$ is known as the **principal root**, a positive value. $-\sqrt{b}$ gives us the negative square root.

In this chapter, we will only focus on problems with numerical radicands. In a future chapter, we will see problems that include variable radicands.

Evaluating Square Roots

Example 1 Evaluate the following.

a) $\sqrt{16}$

b) $-\sqrt{16}$

Solution.

a) $\sqrt{16} = 4$ because $4^2 = 16$.

b) $-\sqrt{16} = -4$ because $-\sqrt{16}$ can be written as $(-1)(\sqrt{16}) = (-1)(4) = -4$.

Exercise 1 **Class Example**
Evaluate the following.

a) $\sqrt{100}$

c) $-\sqrt{49}$

b) $\sqrt{1}$

d) $\sqrt{0.25}$

Exercise 2 **You Try**
Evaluate the following.

a) $\sqrt{64}$

c) $-\sqrt{400}$

b) $\sqrt{0}$

d) $\sqrt{\dfrac{9}{4}}$

Example 2 Evaluate $\sqrt{-25}$.

Solution.
There is no number, a, such that $a^2 = -25$ in the real number system. So, $\sqrt{-25}$ is *not a real number*. The square root of a negative number is not a real number. It is a *complex number*, another branch of mathematics that you will study in a future math class.

Note. Be careful not to confuse $-\sqrt{4}$ and $\sqrt{-4}$. Note that $-\sqrt{4} = -2$ but $\sqrt{-4}$ is not a real number.

Exercise 3 Class Example
Evaluate the following.

a) $\sqrt{-16}$

c) $-\sqrt{4}$

b) $-\sqrt{100}$

d) $-\sqrt{-25}$

Exercise 4 You Try
Evaluate the following.

a) $-\sqrt{121}$

c) $-\sqrt{81}$

b) $\sqrt{-1}$

d) $-\sqrt{-36}$

When the radicand is a perfect square, such as 4, 100, or 0.25, the square root evaluates to a rational number. What about when the radicand is not a perfect square such as 2, 3 or 4.5? The square root is considered an *irrational number*. Numbers such as $\sqrt{2}, \sqrt{4.5}$ are examples of irrational numbers.

Real Numbers

The set of **Real** numbers includes
all **Rational** and **Irrational** numbers

The set of **Rational** numbers consists of
all numbers that can be expressed
as the ratio of two integers

The set of **Irrational** numbers consists of
all numbers that cannot be expressed
as the ratio of two integers

The set of **Integer** numbers consists of

- The number zero

- All *natural numbers*: $\{1,2,3\ldots\}$

- All negatives of natural numbers:
  $\{-1,-2,-3,\ldots\}$

Examples: $\pi, \sqrt{2}, \sqrt{5}, e,$

Examples:

$$\frac{1}{2} = 0.5$$

$$\frac{2}{3} = 0.6\bar{6}$$

The diagram above summarizes different sets of real numbers.
1. The set of **Integers** contains:
    (a) Zero
    (b) Natural Numbers
    (c) Negative values of Natural numbers
    (d) Examples: $-3, -1,; 0, 2, 24$

2. The set of **Rational** numbers contains:

   (a) Numbers that can be represented as a ratio of integers

   (b) Decimals that either terminate or repeat indefinitely

   (c) Examples: $\frac{1}{2} = 0.5$, $\frac{2}{3} = 0.66\overline{6}$, $\frac{5}{1} = 5$

3. The set of **Irrational** numbers contains:

   (a) Cannot be represented as a ratio of integers

   (b) The decimal representations never terminates nor repeats

   (c) Examples: $\pi, \sqrt{2}, -\sqrt{5}, e$

Approximating Irrational Numbers

Since irrational numbers cannot be represented exactly as a decimal, we often leave them in their symbolic form, such as $\sqrt{2}$ or $\sqrt{5}$. If a problem asks for an approximation, we use our calculator to approximate the irrational number.

Example 3 Using a calculator, approximate the following to the nearest hundredth.

a) $\sqrt{2}$
                                            b) $-\sqrt{5}$

Solution.

a) Using a calculator with a 13 decimal place display, $\sqrt{2}$ is shown as 1.4142135623731. Rounding to the nearest hundredth gives us $\sqrt{2} \approx 1.41$. Remember to round correctly. We need to look at the thousandths place or the 3rd decimal place to decide whether to round up or down. (The symbol $\approx$ means "*is approximately*.")

b) Using the same calculator, $-\sqrt{5}$ is shown as -2.23606797749979. Rounding to the nearest hundredth, we get $-\sqrt{5} \approx -2.24$.

Exercise 5 **Class Example**
Using a calculator, approximate the following to the nearest hundredth.

a) $\sqrt{7}$
                                            b) $-\sqrt{11}$

Exercise 6 **You Try**

Using a calculator, approximate the following to the nearest hundredth.

a) $\sqrt{13}$ b) $-\sqrt{3}$

Expressions with Square Roots

Rational and irrational numbers are not alike. So they cannot be combined. For example, $5 + \sqrt{5}$ is the sum of a rational and an irrational number. These expressions cannot be combined in any way. They are unlike terms.

Another example is $4\sqrt{7}$. This expression is a product of a rational and an irrational number. This expression cannot be combined.

However, these expressions can be approximated using a calculator. It is important to save rounding until the very last step. It is best to use your calculator when the expression is in its simplest form and then round at the very end.

Example 4 Using a calculator, approximate $2\sqrt{3}$ to the nearest hundredth.

Solution.

There is a large assortment of calculators and most of them work differently. Be familiar with how your calculator operates. On a calculator that displays 13 decimal places, $2\sqrt{3}$ is shown as 3.46410161513775. Rounding to the nearest hundredth gives us $2\sqrt{3} \approx 3.46$.

Exercise 7 **Class Example**

Using a calculator, approximate the following to 3 decimal places.

a) $6\sqrt{13}$ c) $1 + 7\sqrt{3}$

b) $5 - \sqrt{30}$ d) $\dfrac{8 + \sqrt{6}}{2}$

Exercise 8 Class Example

Using a calculator, approximate the following to 3 decimal places.

a) $-10\sqrt{2}$

c) $\dfrac{2}{3} + \dfrac{\sqrt{5}}{4}$

b) $3 - \sqrt{17}$

d) $\dfrac{2 - \sqrt{14}}{2}$

World View Note The radical sign, when first used was an R with a line through the tail, similar to our prescription symbol, Rx, used today. The R came from the Latin word, "radix", which can be translated as "source" or "foundation." It wasn't until the 1500s that our current symbol was used in Germany. Even then, it was just a check mark, $\sqrt{}$, with no bar over the numbers. In 1637, mathematician, Rene Descartes, first used our present day radical symbol.

7.1: Exercises

Evaluate the following. If the answer is a complex number, state "Not Real."

1. $\sqrt{100}$

2. $-\sqrt{64}$

3. $\sqrt{36}$

4. $-\sqrt{121}$

5. $2\sqrt{1}$

6. $\sqrt{-49}$

7. $3+\sqrt{4}$

8. $7-\sqrt{9}$

9. $\sqrt{\dfrac{25}{16}}$

10. $3\sqrt{\dfrac{4}{9}}$

11. $\dfrac{5+\sqrt{16}}{2}$

12. $\dfrac{1-\sqrt{9}}{3}$

Approximate the following to the nearest tenth. If the answer is a complex number, state "Not Real."

13. $\sqrt{15}$

14. $-\sqrt{-3}$

15. $-\sqrt{21}$

16. $4\sqrt{7}$

17. $-\dfrac{\sqrt{6}}{2}$

18. $4-\sqrt{30}$

19. $10+5\sqrt{3}$

20. $\dfrac{5+\sqrt{3}}{2}$

21. $\dfrac{9-4\sqrt{2}}{5}$

7.2 Simplifying square roots

Objective: To simplify expressions containing square roots

Consider the number $\sqrt{12}$. It is irrational because it cannot be expressed as a ratio of two integers. However, $\sqrt{12}$ can be simplified by determining if the radicand, 12, has any perfect square factors and using the following properties. You might recognize that the properties are very similar to the Product Rule and Quotient Rule for Exponents.

Product Rule for Radicals:

$$\sqrt{ab} = \sqrt{a}\sqrt{b}, \text{ for } a \geqslant 0 \text{ and } b \geqslant 0$$

Quotient Rule for Radicals:

$$\sqrt{\frac{a}{b}} = \frac{\sqrt{a}}{\sqrt{b}}, \text{ for } a \geqslant 0 \text{ and } b > 0$$

Example 1 Simplify $\sqrt{28}$. Give the exact answer and then approximate to 3 decimal places.

Solution.
The question we want to answer is "Does 28 have a factor that is a perfect square?" Yes! We can write $28 = 4 \cdot 7$. Then use the Product Rule for Radicals.

$$\sqrt{28} = \sqrt{4 \cdot 7} \qquad \text{Apply Product Rule for Radicals}$$
$$= \sqrt{4} \cdot \sqrt{7} \qquad \text{Evaluate} \sqrt{4}$$
$$= 2\sqrt{7} \qquad \text{Exact Answer}$$

To approximate, we use a calculator. $2\sqrt{7} \approx 5.292$ rounded to 3 decimal places.

Example 2 Simplify $\sqrt{\frac{3}{25}}$. Give the exact answer and then approximate to the nearest hundredth.

Solution.
We see that the denominator of the fraction is a perfect square and the radicand is a fraction. We

will use the Quotient Rule for Radicals to simplify.

$$\sqrt{\frac{3}{25}} = \frac{\sqrt{3}}{\sqrt{25}}$$ Evaluate $\sqrt{25}$

$$= \frac{\sqrt{3}}{5}$$ Exact Answer

$$\approx 0.35$$ Approximate answer

Note: $\dfrac{\sqrt{3}}{5} = \dfrac{1}{5}\sqrt{3}$

Exercise 1 **Class Example**
Simplify the following. Give the exact answer and then approximate to the nearest hundredth.

a) $\sqrt{8}$ c) $\sqrt{-18}$

b) $-\sqrt{72}$ d) $\sqrt{\dfrac{50}{9}}$

Exercise 2 **You Try**
Simplify the following. Give the exact answer and then approximate to 3 decimal places.

a) $\sqrt{24}$ c) $\sqrt{-24}$

b) $-\sqrt{200}$ d) $\sqrt{\dfrac{2}{49}}$

Example 3 Simplify $\dfrac{8-\sqrt{12}}{2}$. Give the exact answer and then approximate to 2 decimal places.

Solution.
First simplify $\sqrt{12}$ in the numerator. We can rewrite $12 = 4 \cdot 3$ where one of the factors is a perfect square.

$$
\begin{aligned}
\sqrt{12} &= \sqrt{4 \cdot 3} && \text{Apply Product Rule for Radicals} \\
&= \sqrt{4} \cdot \sqrt{3} && \text{Evaluate } \sqrt{4} \\
&= 2\sqrt{3}
\end{aligned}
$$

Substitute this result, $\sqrt{12} = 2\sqrt{3}$ into the original problem and simplify.

$$
\begin{aligned}
\frac{8-\sqrt{12}}{2} &= \frac{8-2\sqrt{3}}{2} && \text{Factor out GCF in numerator} \\
&= \frac{2(4-\sqrt{3})}{2} && \text{Simplify} \\
&= 4-\sqrt{3} && \text{Exact Answer} \\
&\approx 2.27 && \text{Approximate Answer}
\end{aligned}
$$

Note. There is an alternate approach we can use to simplify $\dfrac{8-2\sqrt{3}}{2}$. First rewrite the expression as two separate fractions. This technique can be applied only when the denominator is a monomial.

$$
\begin{aligned}
\frac{8-2\sqrt{3}}{2} &= \frac{8}{2} - \frac{2\sqrt{3}}{2} \\
&= 4 - \sqrt{3}
\end{aligned}
$$

Exercise 3 **Class Example**
Simplify $\dfrac{6+\sqrt{8}}{10}$. Give the exact answer and then approximate to the nearest tenth.

Exercise 4 **You Try**

Simplify the following. Give the exact answer and then approximate to 1 decimal place.

a) $\dfrac{3-\sqrt{72}}{3}$

b) $\dfrac{10+\sqrt{20}}{14}$

Squaring a Square Root Term

What happens when we square a square root term such as $(\sqrt{7})^2$? Let us take a look.

Example 4 Simplify $(\sqrt{7})^2$.

Solution.

$$
\begin{aligned}
(\sqrt{7})^2 &= (\sqrt{7}) \cdot (\sqrt{7}) && \text{Expand} \\
&= \sqrt{7 \cdot 7} && \text{Apply Product Rule for Radicals} \\
&= \sqrt{49} && \text{Evaluate} \\
&= 7 && \text{Our Solution}
\end{aligned}
$$

From the above example, we see that if we square a square root term, we get back the radicand. This leads us to the following property.

$$
\left(\sqrt{b}\right)^2 = b \text{ when } b \geqslant 0
$$

To square a square root term when there is a number in front of the square root, we apply the Power of a Product Rule for Exponents. This leads us to the following general property for squaring a square root term.

$$
\left(a\sqrt{b}\right)^2 = (a)^2 \left(\sqrt{b}\right)^2 = a^2 \cdot b \text{ when } b \geqslant 0
$$

Example 5 Simplify $(-4\sqrt{7})^2$.

Solution.

$$\begin{aligned}(-4\sqrt{7})^2 &= (-4)^2(\sqrt{7})^2 \qquad &&\text{Apply property for squaring a square root term}\\ &= (16)(7) &&\text{Evaluate}\\ &= 112 &&\text{Our Solution}\end{aligned}$$

Exercise 5 Class Example
Simplify the following.

a) $\left(\sqrt{6}\right)^2$

c) $\left(3\sqrt{5}\right)^2$

b) $\left(-\sqrt{12}\right)^2$

d) $\left(-4\sqrt{2}\right)^2$

Exercise 6 You Try
Simplify the following.

a) $\left(\sqrt{11}\right)^2$

c) $\left(5\sqrt{3}\right)^2$

b) $\left(-\sqrt{8}\right)^2$

d) $\left(-2\sqrt{6}\right)^2$

7.2: Exercises

Simplify the following. Give an exact answer, and if irrational, approximate to the nearest hundredth. If the answer is a complex number, state "Not Real."

1. $\sqrt{8}$

2. $\sqrt{32}$

3. $-\sqrt{45}$

4. $\sqrt{-20}$

5. $\dfrac{\sqrt{72}}{4}$

6. $\sqrt{\dfrac{20}{9}}$

7. $3 + \sqrt{50}$

8. $7 - \sqrt{12}$

9. $\dfrac{1}{2} + \dfrac{\sqrt{63}}{2}$

10. $\dfrac{6 + \sqrt{28}}{2}$

11. $\dfrac{5 + \sqrt{24}}{5}$

12. $\dfrac{15 - \sqrt{90}}{6}$

Chapter 7 Assessment

Simplify the following expressions.

1. $\sqrt{81}$

2. $3\sqrt{36}$

3. $\sqrt{\dfrac{25}{64}}$

4. $\dfrac{4\sqrt{100}}{30}$

5. $12 - \sqrt{16}$

6. $\dfrac{9 - \sqrt{49}}{4}$

Approximate the following expressions to the nearest hundredth.

7. $\sqrt{22}$

8. $-\sqrt{183}$

9. $\sqrt{\dfrac{2}{5}}$

10. $\dfrac{19 + \sqrt{5}}{2}$

Simplify the following. Give exact answers and if irrational, approximate to the nearest hundredth.

12. $\sqrt{12}$

13. $3\sqrt{32}$

14. $\sqrt{\dfrac{11}{25}}$

15. $\dfrac{\sqrt{50}}{5}$

16. $-\sqrt{80}$

17. $\dfrac{6 - \sqrt{48}}{4}$

8. Graphs of Quadratic Equations

8.1 Square Root Property

Objective: To solve quadratic equations using the square root property

Square Root Property

In this section, we will use the *Square Root Property* to find solutions to quadratic equations. To help motivate this property, we first make a few observations. Consider the following two equations.

$$\sqrt{(3)^2} = \sqrt{9} = 3$$

$$\sqrt{(-3)^2} = \sqrt{9} = 3$$

More generally, what can we say about $\sqrt{x^2}$? From the above two equations, we observe the following.

- When x is a positive real number, $\sqrt{x^2}$ is equal to x itself. This happened in the first equation above, $\sqrt{(3)^2} = 3$, since $x = 3$, a positive real number.

- However, when x is a negative real number, $\sqrt{x^2}$ is equal to $-x$. This happened in the second equation above, $\sqrt{(-3)^2} = -(-3) = 3$ since $x = -3$.

This means that taking the square root of a number, after it was squared, $\sqrt{x^2}$, has the same effect as taking the absolute value of the number. This is known as the *Square Root Property*.

Square Root Property:

$$\sqrt{x^2} = |x|$$

Expressions, such as $\sqrt{(x+5)^2}$, still use the square root property. That is, $\sqrt{(x+5)^2} = |x+5|$.

Example 1 Simplify the following expressions.

a) $\sqrt{3^2}$

c) $\sqrt{z^2}$

b) $\sqrt{(-3)^2}$

d) $\sqrt{(y-7)^2}$

Solution.
We will simplify each of these by using the Square Root Property.

a) $\sqrt{3^2} = |3| = 3$

b) $\sqrt{(-3)^2} = |-3| = 3$

c) $\sqrt{z^2} = |z|$

d) $\sqrt{(y-7)^2} = |y-7|$

Exercise 1 Class Example
Simplify the following expressions.

a) $\sqrt{4^2}$

c) $\sqrt{(5x)^2}$

b) $\sqrt{(-4)^2}$

d) $\sqrt{(2m+3)^2}$

Exercise 2 You Try
Simplify the following expressions.

a) $\sqrt{5^2}$

c) $\sqrt{p^2}$

b) $\sqrt{(-5)^2}$

d) $\sqrt{(7-3w)^2}$

Quadratic Equations

We are now in a position to solve quadratic equations using the Square Root Property. This will then lead to an absolute value equation which we can solve using techniques we learned in Chapter 1.

Example 2 Solve the following equations. Be sure to simplify your answer and check your solution. If the answer is irrational, round to the nearest hundredth.

a) $x^2 = 4$

b) $x^2 = 7$

Solution.

a)

$$x^2 = 4 \qquad \text{Take the square root of each side}$$
$$\sqrt{x^2} = \sqrt{4} \qquad \text{Simplify}$$
$$|x| = 2 \qquad \text{Solve absolute value equations}$$
$$x = 2 \text{ or } x = -2 \qquad \text{Our Solution}$$

Verify that we have the correct solution by substituting our answer into the original quadratic equation, $x^2 = 4$.

When $x = 2$:

$$(2)^2 = 4 \ \checkmark$$

When $x = -2$:

$$(-2)^2 = 4 \ \checkmark$$

Our solution to the quadratic equation is $x = 2$ or $x = -2$.

b)

$$x^2 = 7 \qquad \text{Take the square root of each side}$$
$$\sqrt{x^2} = \sqrt{7} \qquad \text{Simplify}$$
$$|x| = \sqrt{7} \qquad \text{Solve absolute value equation}$$
$$x = \sqrt{7} \text{ or } x = -\sqrt{7} \qquad \text{Exact Solution}$$
$$x \approx 2.65 \text{ or } x \approx -2.65 \qquad \text{Approximate Solution}$$

Verify that we have the correct solution by substituting our answer into the original quadratic equation, $x^2 = 7$.

When $x = \sqrt{7}$: $\qquad\qquad\qquad\qquad$ When $x = -\sqrt{7}$:
$$(\sqrt{7})^2 = 7 \ \checkmark \qquad\qquad\qquad (-\sqrt{7})^2 = 7 \ \checkmark$$

Our solution to the quadratic equation are $x = \sqrt{7}$ or $x = -\sqrt{7}$.

There are several key points to take note of from the above examples.

1. We can solve the equation, $x^2 = 4$, by factoring. That is because 4 is a perfect square.

$$x^2 = 4 \qquad \text{Write the equation in standard form}$$
$$x^2 - 4 = 0 \qquad \text{Factor left side of equation}$$
$$(x+2)(x-2) = 0 \qquad \text{Use zero product property}$$
$$x+2 = 0 \text{ or } x-2 = 0 \qquad \text{Solve for } x$$
$$x = -2 \text{ or } x = 2 \qquad \text{Our Solution}$$

2. However, we won't be able to solve the equation, $x^2 = 7$ by factoring. Rewriting the quadratic equation in standard form gives us:

$$x^2 - 7 = 0$$

The polynomial, $x^2 - 7$ is prime. Therefore, it is not factorable. We need to use the property of square roots to solve the quadratic equation.

3. We can condense our solutions using what is known as the "plus and minus" notation, that is, $\pm$. For example, instead of writing:

$$x = 2 \ \text{ or } \ x = -2$$

we can write:

$$x = \pm 2$$

It is important to remember that when writing the expression $x = \pm\sqrt{7}$, you are really specifying two values of x, namely, $x = \sqrt{7}$ or $x = -\sqrt{7}$.

4. When checking our answers, use the exact form (ie, the simplified radical form) rather than the approximate form (ie, the number that was rounded from the calculator output) to avoid discrepancies. For example, if we had used $x \approx 2.65$ instead of $x = \sqrt{7}$ to check our answer for the equation $x^2 = 7$, we would have gotten the following.

$$x^2 = 7$$

$$(2.65)^2 \overset{?}{=} 7$$

$$7.0025 \neq 7$$

Since $\sqrt{7}$ was approximated to 2.65, we will not get exactly 7 when 2.65 is squared, although the answer, 7.0025 is close to 7. To get exactly 7, we will need to use our exact solution, $x = \sqrt{7}$, to check our answer.

5. More generally, we have the following useful property which we will use from now on.

> If $|A| = B$, then $A = \pm B$

Example 3 Solve $x^2 = 18$. Be sure to simplify your answer and check your solution. If the answer is irrational, round to the nearest hundredth.

Solution.

$x^2 = 18$	Take the square root of each side
$\sqrt{x^2} = \sqrt{18}$	Simplify
$\|x\| = 3\sqrt{2}$	Solve absolute value equation
$x = \pm 3\sqrt{2}$	Exact Solution
$x \approx \pm 4.24$	Approximate Solution

Verify that we have the correct solution by substituting our answer into the original quadratic equation, $x^2 = 18$.

When $x = 3\sqrt{2}$:

$$(3\sqrt{2})^2 \overset{?}{=} 18$$

$$(3)^2(\sqrt{2})^2 \overset{?}{=} 18$$

$$9(2) = 18 \ \checkmark$$

When $x = -3\sqrt{2}$:

$$(-3\sqrt{2})^2 \overset{?}{=} 18$$

$$(-3)^2(\sqrt{2})^2 \overset{?}{=} 18$$

$$9(2) = 18 \ \checkmark$$

Our solution to the quadratic equation are $x = 3\sqrt{2}$ or $x = -3\sqrt{2}$.

Example 4 Solve $(x-3)^2 = 11$. Be sure to simplify your answer and check your solution. If the answer is irrational, round to the nearest tenth.

Solution.

$$(x-3)^2 = 11$$ Take the square root of each side

$$\sqrt{(x-3)^2} = \sqrt{11}$$ Simplify

$$|x-3| = \sqrt{11}$$ Solve absolute value equation

$$x-3 = \pm\sqrt{11}$$ Add 3 to each side

$$x = 3 \pm \sqrt{11}$$ Exact Solution

$$x \approx 6.3 \text{ or } x \approx -0.3$$ Approximate Solution

Verify that we have the correct solution by substituting our answer into the original quadratic equation, $(x-3)^2 = 11$.

When $x = 3 + \sqrt{11}$:   When $x = 3 - \sqrt{11}$:

$$[(3+\sqrt{11})-3]^2 \overset{?}{=} 11$$ $$[(3-\sqrt{11})-3]^2 \overset{?}{=} 11$$

$$[3+\sqrt{11}-3]^2 \overset{?}{=} 11$$ $$[3-\sqrt{11}-3]^2 \overset{?}{=} 11$$

$$[\sqrt{11}]^2 = 11 \checkmark$$ $$[\sqrt{11}]^2 = 11 \checkmark$$

Our solution to the quadratic equation is $x = 3 + \sqrt{11}$ or $x = 3 - \sqrt{11}$.

Example 5 Solve $(x+4)^2 = 9$. Be sure to simplify your answer and check your solution. If the answer is irrational, round to the nearest hundredth.

Solution.

$$(x+4)^2 = 9$$ Take the square root of each side

$$\sqrt{(x+4)^2} = \sqrt{9}$$ Simplify

$$|x+4| = 3$$ Solve absolute value equation

$$x+4 = \pm 3$$ Subtract 4 from each side

$$x = -4 \pm 3$$ Perform indicated operation

$$x = -4+3 \text{ or } x = -4-3$$ Simplify

$$x = -1 \text{ or } x = -7$$ Our Solution

Verify that we have the correct solution by substituting our answer into the original quadratic

equation, $(x+4)^2 = 9$.

When $x = -1$:

$$(-1+4)^2 \stackrel{?}{=} 9$$
$$(3)^2 = 9 \checkmark$$

When $x = -7$:

$$(-7+4)^2 \stackrel{?}{=} 9$$
$$(-3)^2 = 9 \checkmark$$

Our solution to the quadratic equation is $x = -1$ or $x = -7$.

Example 6 Solve $(2x-1)^2 = 7$. Be sure to simplify your answer and check your solution. If the answer is irrational, round to the nearest hundredth.

Solution.

$$(2x-1)^2 = 7 \qquad \text{Take the square root of each side}$$

$$\sqrt{(2x-1)^2} = \sqrt{7} \qquad \text{Simplify}$$

$$|2x-1| = \sqrt{7} \qquad \text{Solve absolute value equation}$$

$$2x-1 = \pm\sqrt{7} \qquad \text{Add 1 to each side}$$

$$2x = 1 \pm \sqrt{7} \qquad \text{Divide each side by 2}$$

$$x = \frac{1 \pm \sqrt{7}}{2} \qquad \text{Exact Solution}$$

$$x \approx 1.82 \text{ or } x \approx -0.82 \qquad \text{Approximate Solution}$$

Verify that we have the correct solution by substituting our answer into the original quadratic equation, $(2x-1)^2 = 7$.

When $x = \dfrac{1+\sqrt{7}}{2}$:

$$\left[2\left(\frac{1+\sqrt{7}}{2}\right) - 1\right]^2 \stackrel{?}{=} 7$$

$$[(1+\sqrt{7})-1]^2 \stackrel{?}{=} 7$$

$$[1+\sqrt{7}-1]^2 \stackrel{?}{=} 7$$

$$[\sqrt{7}]^2 = 7 \checkmark$$

When $x = \dfrac{1-\sqrt{7}}{2}$:

$$\left[2\left(\frac{1-\sqrt{7}}{2}\right) - 1\right]^2 \stackrel{?}{=} 7$$

$$[(1-\sqrt{7})-1]^2 \stackrel{?}{=} 7$$

$$[1-\sqrt{7}-1]^2 \stackrel{?}{=} 7$$

$$[-\sqrt{7}]^2 = 7 \checkmark$$

Our solution to the quadratic equation is $x = \dfrac{1+\sqrt{7}}{2}$ or $x = \dfrac{1-\sqrt{7}}{2}$.

Example 7 Solve $(x+6)^2 = -4$. Be sure to simplify your answer and check your solution. If the answer is irrational, round to the nearest hundredth.

Solution.

$$(x+6)^2 = -4 \qquad\qquad \text{Take square root of each side}$$

$$\sqrt{(x+6)^2} = \sqrt{-4} \qquad\qquad \sqrt{-4} \text{ is not a real number!}$$

This equation does not have any real number solutions. The solutions are complex numbers.

Exercise 3 Class Example

Solve the following equations. Be sure to simplify your answer and check your solution. If the answer is irrational, round to the nearest thousandth.

a) $m^2 = \dfrac{2}{9}$

c) $(3x+2)^2 = 36$

b) $(y-5)^2 = 24$

d) $(g+8)^2 = -16$

Exercise 4 You Try

Solve the following equations. Be sure to simplify your answer and check your solution. If the answer is irrational, round to the nearest hundredth.

a) $x^2 = \dfrac{3}{4}$

c) $(p-2)^2 = -8$

b) $(y+7)^2 = 49$

d) $(2w+4)^2 = 12$

When solving quadratic equations using the square root property, we need to first isolate the square term, and then apply the Square Root Property. The following examples show how this is done.

Example 8 Solve $2x^2 - 1 = 17$. Be sure to simplify your answer and check your solution. If the answer is irrational, round to the nearest thousandth.

Solution.

$2x^2 - 1 = 17$	Isolate the x^2 term by adding 1 to each side
$2x^2 = 18$	Divide each side by 2
$x^2 = 9$	Take the square root of each side
$\sqrt{x^2} = \sqrt{9}$	Simplify
$|x| = 3$	Solve absolute value equation
$x = \pm 3$	Our Solution

Verify that we have the correct solution by substituting our answer into the original quadratic equation, $2x^2 - 1 = 17$.

When $x = 3$:

$$2(3)^2 - 1 \overset{?}{=} 17$$

$$2(9) - 1 \overset{?}{=} 17$$

$$18 - 1 = 17 \checkmark$$

When $x = -3$:

$$2(-3)^2 \overset{?}{=} 17$$

$$2(9) - 1 \overset{?}{=} 17$$

$$18 - 1 = 17 \checkmark$$

Our solution to the quadratic equation is $x = 3$ or $x = -3$.

Example 9 Solve $(x-4)^2 - 3 = 5$. Be sure to simplify your answer and check your solution. If the answer is irrational, round to the nearest hundredth.

Solution.

$(x-4)^2 - 3 = -5$	Isolate $(x-4)^2$ term by adding 3 to each side
$(x-4)^2 = 8$	Take the square root of each side
$\sqrt{(x-4)^2} = \sqrt{8}$	Simplify
$|x-4| = 2\sqrt{2}$	Solve absolute value equation
$x - 4 = \pm 2\sqrt{2}$	Add 4 to each side
$x = 4 \pm 2\sqrt{2}$	Exact Solution
$x \approx 6.83$ or $x \approx 1.17$	Approximate Solution

Verify that we have the correct solution by substituting our answer into the original quadratic

equation, $(x-4)^2 - 3 = 5$.

When $x = 4 + 2\sqrt{2}$:

$$[(4+2\sqrt{2})-4]^2 - 3 \stackrel{?}{=} 5$$

$$[4+2\sqrt{2}-4]^2 - 3 \stackrel{?}{=} 5$$

$$[2\sqrt{2}]^2 - 3 \stackrel{?}{=} 5$$

$$(2)^2(\sqrt{2})^2 - 3 \stackrel{?}{=} 5$$

$$4(2) - 3 \stackrel{?}{=} 5$$

$$8 - 3 = 5 \quad \checkmark$$

When $x = 4 - 2\sqrt{2}$:

$$[(4-2\sqrt{2})-4]^2 - 3 \stackrel{?}{=} 5$$

$$[4-2\sqrt{2}-4]^2 - 3 \stackrel{?}{=} 5$$

$$[-2\sqrt{2}]^2 - 3 \stackrel{?}{=} 5$$

$$(-2)^2(\sqrt{2})^2 \stackrel{?}{=} 5$$

$$4(2) - 3 \stackrel{?}{=} 5$$

$$8 - 3 = 5 \quad \checkmark$$

Our solution to the quadratic equation is $x = 4 + 2\sqrt{2}$ or $x = 4 - 2\sqrt{2}$.

Example 10 Solve $(2x-5)^2 + 7 = 4$. Be sure to simplify your answer and check your solution. If the answer is irrational, round to the nearest hundredth.

Solution.

$$(2x-5)^2 + 7 = 4 \qquad \text{Isolate the } (2x-5)^2 \text{ term by subtracting 7 from each side}$$

$$(2x-5)^2 = -3 \qquad \text{Take square root of each side}$$

$$\sqrt{(2x-5)^2} = \sqrt{-3} \qquad \sqrt{-3} \text{ is not a real number!}$$

This equation does not have any real number solutions. The solutions are complex numbers.

Example 11 Solve $4(6x+5)^2 - 3 = 33$. Be sure to simplify your answer and check your solution. If the answer is irrational, round to the nearest tenth.

Solution.

$4(6x+5)^2 - 3 = 33$	Isolate $(6x+5)^2$ term by adding 3 to each side
$4(6x+5)^2 = 36$	Divide each side by 4
$(6x+5)^2 = 9$	Take the square root of each side
$\sqrt{(6x+5)^2} = \sqrt{9}$	Simplify
$|6x+5| = 3$	Solve absolute value equation
$6x+5 = \pm 3$	Subtract 5 to each side
$6x = -5 \pm 3$	Divide each side by 6
$x = \dfrac{-5 \pm 3}{6}$	Perform indicated operataion
$x = \dfrac{-5+3}{6}$ or $x = \dfrac{-5-3}{6}$	Perform indicated operation
$x = \dfrac{-2}{6}$ or $x = \dfrac{-8}{6}$	Simplify
$x = -\dfrac{1}{3}$ or $x = -\dfrac{4}{3}$	Our Solution

Verify that we have the correct solution by substituting our answer into the original quadratic equation, $4(6x+5)^2 - 3 = 33$.

When $x = -\dfrac{1}{3}$:

$$4\left[6\left(-\dfrac{1}{3}\right)+5\right]^2 - 3 \overset{?}{=} 33$$

$$4[-2+5]^2 - 3 \overset{?}{=} 33$$

$$4[\,3\,]^2 - 3 \overset{?}{=} 33$$

$$4(9) - 3 \overset{?}{=} 33$$

$$36 - 3 = 33 \;\checkmark$$

When $x = -\dfrac{4}{3}$:

$$4\left[6\left(-\dfrac{4}{3}\right)+5\right]^2 - 3 \overset{?}{=} 33$$

$$4[-8+5]^2 - 3 \overset{?}{=} 33$$

$$4[-3]^2 - 3 \overset{?}{=} 33$$

$$4(9) - 3 \overset{?}{=} 33$$

$$36 - 3 = 33 \;\checkmark$$

Our solution to the quadratic equation is $x = -\dfrac{1}{3}$ or $x = -\dfrac{4}{3}$.

Exercise 5 Class Example
Solve the following equations. Be sure to simplify your answer and check your solution. If the answer is irrational, round to the nearest tenth.

a) $4y^2 - 1 = 6$

c) $(2c - 1)^2 - 5 = 7$

b) $(x - 7)^2 - 3 = 6$

d) $3 - 5(3p + 4)^2 = 13$

Exercise 6 **You Try**

Solve the following equations. Be sure to simplify your answer and check your solution. If the answer is irrational, round to the nearest thousandth.

a) $9w^2 + 3 = 7$

c) $5(2w-1)^2 = 15$

b) $(y+3)^2 + 4 = 9$

d) $1 - 3(x+6)^2 = 4$

8.1: Exercises

Fill in the blanks.

1. $\sqrt{\boxed{}^2} = $ _____ is known as the _____ Property.

Simplify the following expressions.

2. $\sqrt{(7)^2}$

3. $\sqrt{(-7)^2}$

4. $\sqrt{A^2}$

5. $\sqrt{(A+B)^2}$

6. $\sqrt{(x+3)^2}$

7. $\sqrt{(2y-9)^2}$

8. $\sqrt{x^2+6x+9}$

Solve each eqeuation. Be sure to fully simplilfy and check each of your solutions using the exact form of your answer. If answers are irrational, round to the nearest hundredth.

9. $x^2 = 36$

10. $y^2 = 11$

11. $z^2 = 32$

12. $w^2 = -4$

13. $x^2 = 13$

14. $x^2 + 4 = 9$

15. $3y^2 = 18$

16. $2y^2 = 144$

17. $3z^2 - 29 = -5$

18. $\dfrac{1}{10}v^2 = 30$

19. $5 - 2t^2 = 7t^2 + 4$

20. $(x-3)^2 = 7$

21. $(z-1)^2 = 2$

22. $(y+4)^2 + 3 = 52$

23. $4 - (h+2)^2 = 0$

24. $(2t+5)^2 = 10$

25. $(4y+8)^2 = 12$

26. $(3g-6)^2 - 18 = 0$

27. $8 - (2p+1)^2 = -1$

28. $100t^2 - 400 = 2000$

29. $5 - 3x^2 = 6 - 2x^2$

30. $5(x-8)^2 = 25$

31. $2(3y+4)^2 - 7 = 9$

8.2 Completing the square

Objective: To solve quadratic equations by completing the square

We will learn another method for solving quadratic equations known as completing the square. The idea is to change the quadratic expression into a perfect square trinomial and then use the square root property to solve for the variable. Let us first take a look at an example at how this works.

Example 1 Solve $x^2 + 2x + 1 = 7$. Leave the answer in exact and simplified form.

Solution.
The idea behind solving a quadratic by completing the square is to have a perfect square trinomial. Recall from the previous chapter how we can recognize a perfect square trinomial. Notice that the quadratic term, x^2, and the constant term, 1, are perfect squares. The square root of the quadratic term is x and the square root of the constant term is 1. When the quadratic term and constant term are multiplied together, the product is $1x$ or x. When x is multiplied by 2, we get the linear term, $2x$, which is also the linear term of the trinomial. This indicates that the trinomial, $x^2 + 2x + 1$, is a perfect square trinomial and factors as $(x+1)^2$. We can use the method from the previous section to solve the quadratic equation.

$$x^2 + 2x + 1 = 7 \qquad \text{Factor perfect square trinomial}$$
$$(x+1)^2 = 7 \qquad \text{Take square root of each side}$$
$$\sqrt{(x+1)^2} = \sqrt{7} \qquad \text{Simplify}$$
$$|x+1| = \sqrt{7} \qquad \text{Solve absolute value equation}$$
$$x+1 = \pm\sqrt{7} \qquad \text{Subtract 1 from each side}$$
$$x = -1 \pm \sqrt{7} \qquad \text{Exact and Simplified Solution}$$

The trinomial in the above example is already a perfect square trinomial. Unfortunately, that situation does not occur often. Our goal then is to get the trinomial to become a perfect square trinomial.

Forming a Perfect Square Trinomial Algebraically

We will first practice changing quadratic expressions of the form $x^2 + bx$ into a perfect square trinomial. To do so, we will be searching for the constant term, c, to add to the quadratic expression in order to make it a perfect square trinomial. This is done by squaring half of the linear term, that is $c = \left(\frac{1}{2} \cdot b\right)^2$.

Let's see where this comes from. Suppose we have:

$$(x+p)^2 = x^2 + 2px + p^2$$

The trinomial is similar in form to the general quadratic expression, $x^2 + bx + c$, where $2p = b$ and

$p^2 = c$. Solving for p, we get:

$$2p = b \qquad\qquad\qquad\qquad \text{Divide each side by 2}$$

$$p = \frac{1}{2} \cdot b \qquad\qquad\qquad \text{Square each side}$$

$$p^2 = \left(\frac{1}{2} \cdot b\right)^2 \qquad\qquad \text{Substitute } p^2 \text{ with } c$$

$$c = \left(\frac{1}{2} \cdot b\right)^2 \qquad\qquad \text{Our Formula}$$

Forming a Perfect Square Trinomial Geometrically

Let us look at how the constant term, c, is derived from a geometric standpoint. Our goal is to change the expression $x^2 + bx$ into a perfect square trinomial by completing the square.

A figure whose area is x^2 means its length and width are both x. Only a square can have equal length and width. A figure whose area is bx means that one side is of length, b, and another side is of length, x. This figure must be a rectangle.

Next, cut side, b in half to get $\frac{b}{2}$ or $\frac{1}{2}b$. Connect the ends as follows.

$$= x^2 + \frac{b}{2}x + \frac{b}{2}x$$
$$= x^2 + bx$$

We now need to *complete the square* of the above figure by finding the area of the missing square in question.

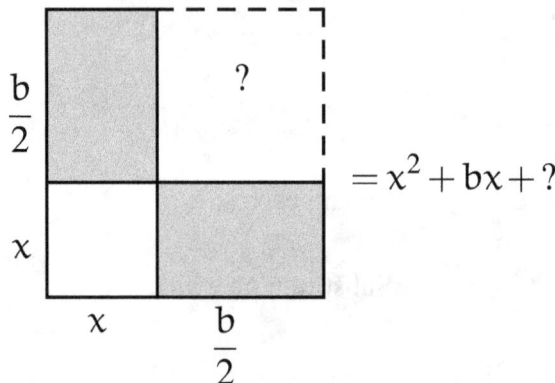

$$= x^2 + bx + ?$$

The missing square in question has sides of length, $\frac{b}{2} = \frac{1}{2}b$. Therefore, the area of the missing square is $\left(\frac{1}{2}b\right)\left(\frac{1}{2}b\right) = \left(\frac{1}{2}\,b\right)^2$. This is the constant term, c, that needs to be added to the expression, $x^2 + bx$, in order to form a perfect square trinomial and to *complete the square* of the figure. Therefore, $c = \left(\frac{1}{2}\,b\right)^2$.

Example 2 Find the constant term, c, to make $x^2 + 8x + c$ into a perfect square trinomial. Then factor the trinomial.

Solution.
To find the constant term, c, first identify b, the coefficient of x. In this case, $b = 8$.

$$c = \left(\frac{1}{2}b\right)^2 \qquad\qquad \text{Substitute } b = 8$$

$$= \left(\frac{1}{2}\cdot 8\right)^2 \qquad\qquad \text{Multiply}$$

$$= (4)^2 \qquad\qquad \text{Square}$$

$$= 16$$

Substitute $c = 16$ into the trinomial and factor.

$$x^2 + 8x + c \qquad\qquad \text{Substitute } c = 16$$
$$x^2 + 8x + 16 \qquad\qquad \text{Factor perfect square trinomial}$$
$$= (x + 4)^2 \qquad\qquad \text{Our factor}$$

Check to verify factor is correct.

$$(x+4)^2 = (x+4)(x+4)$$
$$= x(x+4) + 4(x+4)$$
$$= x^2 + 4x + 4x + 16$$
$$= x^2 + 8x + 16 \quad \checkmark$$

Let us see how this problem can be done geometrically.

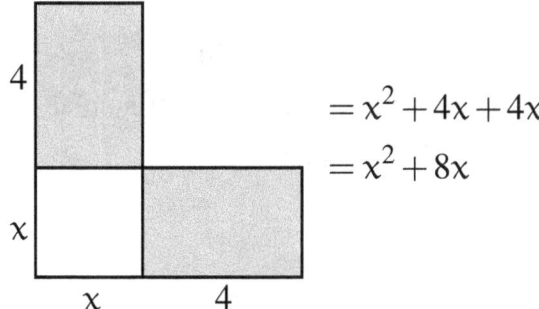

Cut the side with length 8 units in half and connect the ends as follows.

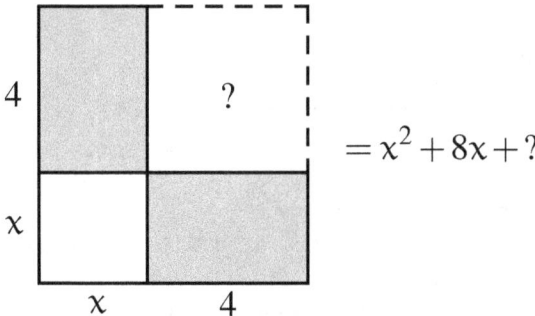

$$= x^2 + 4x + 4x$$
$$= x^2 + 8x$$

We now need to *complete the square* of the above figure by finding the area of the missing square in question.

$$= x^2 + 8x + ?$$

The missing square in question has sides of length 4. Its area is $4 \cdot 4 = 16$. Therefore, 16 must be added to $x^2 + 8x$ in order to form a perfect square trinomial and to complete the square of the figure.

Example 3 Find the constant term, c, to make $x^2 - 16x + c$ into a perfect square trinomial. Then factor the trinomial.

Solution.
To find the constant term, c, first identify b, the coefficient of x. In this case, $b = -16$.

$$c = \left(\frac{1}{2}b\right)^2 \qquad\qquad \text{Substitute } b = -16$$

$$= \left(\frac{1}{2} \cdot -16\right)^2 \qquad\qquad \text{Multiply}$$

$$= (-8)^2 \qquad\qquad \text{Square}$$

$$= 64$$

Substitute $c = 64$ into the trinomial and factor.

$$x^2 - 16x + c \qquad\qquad \text{Substitute } c = -64$$
$$x^2 - 16x + 64 \qquad\qquad \text{Factor perfect square trinomial}$$
$$(x - 8)^2 \qquad\qquad \text{Our Solution}$$

Check to verify factor is correct.

$$(x - 8)^2 = (x - 8)(x - 8)$$
$$= x(x - 8) - 8(x - 8)$$
$$= x^2 - 8x - 8x + 64$$
$$= x^2 - 16x + 64 \checkmark$$

Let us see how this problem can be done geometrically.

Cut the side with length "-16" units in half and connect the ends as follows.

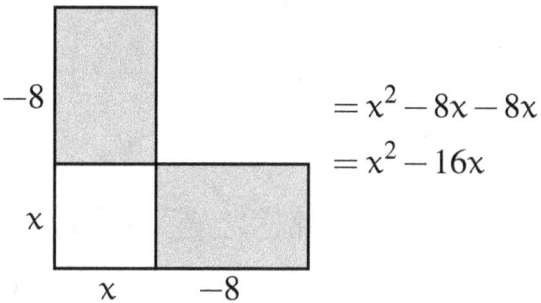

$$= x^2 - 8x - 8x$$
$$= x^2 - 16x$$

We now need to complete the square of the above figure by finding the area of the missing square in question.

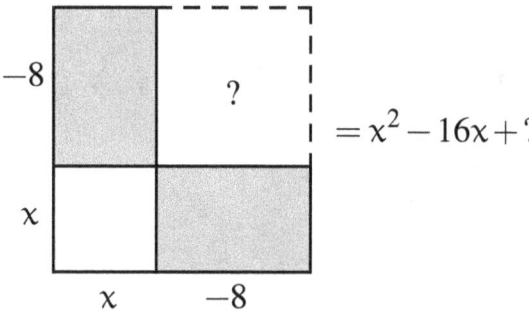

$$= x^2 - 16x + ?$$

The missing square in question has sides of length "−8". Its area is $(-8) \cdot (-8) = 64$. Therefore, 64 must be added to $x^2 - 16x$ in order to form a perfect square trinomial and to complete the square of the figure.

Example 4 Find the constant term, c, to make $y^2 + 7y + c$ into a perfect square trinomial. Then factor the trinomial.

Solution.
To find the constant term, c, first identify b, the coefficient of x. In this case, $b = 7$.

$$c = \left(\frac{1}{2}b\right)^2 \qquad\qquad \text{Substitute } b = 7$$

$$= \left(\frac{1}{2} \cdot 7\right)^2 \qquad\qquad \text{Multiply}$$

$$= \left(\frac{7}{2}\right)^2 \qquad\qquad \text{Square}$$

$$= \frac{49}{4}$$

Substitute $c = \dfrac{49}{4}$ into the trinomial and factor.

$$y^2 + 7y + c \qquad\qquad\qquad \text{Substitute } c = \dfrac{49}{4}$$

$$y^2 + 7y + \dfrac{49}{4} \qquad\qquad\qquad \text{Factor perfect square trinomial}$$

$$\left(y + \dfrac{7}{2}\right)^2 \qquad\qquad\qquad \text{Our Solution}$$

Check to verify factor is correct.

$$\left(y + \dfrac{7}{2}\right)^2 = \left(y + \dfrac{7}{2}\right)\left(y + \dfrac{7}{2}\right)$$

$$= y\left(y + \dfrac{7}{2}\right) + \dfrac{7}{2}\left(y + \dfrac{7}{2}\right)$$

$$= y^2 + \dfrac{7}{2}y + \dfrac{7}{2}y + \dfrac{49}{4}$$

$$= y^2 + \dfrac{7}{2}y + \dfrac{49}{4} \quad \checkmark$$

Let us see how this problem can be done geometrically.

Cut the side with length 7 units in half and connect the ends as follows.

$$= x^2 + \dfrac{7}{2}x + \dfrac{7}{2}x$$

$$= x^2 + 7x$$

We now need to complete the square of the above figure by finding the area of the missing square in question.

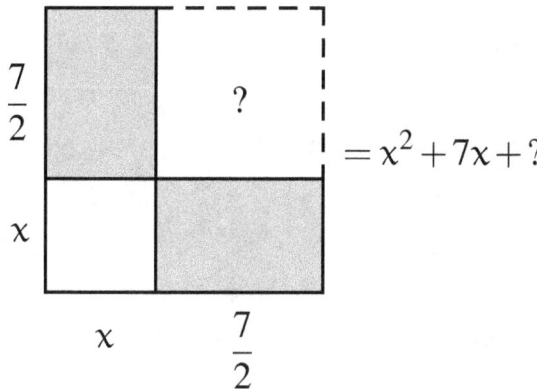

$$= x^2 + 7x + ?$$

The missing square in question has sides of length $\frac{7}{2}$. Its area is $\left(\frac{7}{2}\right) \cdot \left(\frac{7}{2}\right) = \frac{49}{4}$. Therefore, $\frac{49}{4}$ must be added to $y^2 + 7y$ in order to form a perfect square trinomial and to complete the square of the figure.

Example 5 Find the constant term, c, to make $y^2 - \frac{5}{3}y + c$ into a perfect square trinomial. Then factor the trinomial.

Solution.
To find the constant term, c, first identify b, the coefficient of x. In this case, $b = -\frac{5}{3}$.

$$c = \left(\frac{1}{2}b\right)^2 \qquad\qquad \text{Substitute } b = -\frac{5}{3}$$

$$= \left(\frac{1}{2} \cdot -\frac{5}{3}\right)^2 \qquad\qquad \text{Multiply}$$

$$= \left(-\frac{5}{6}\right)^2 \qquad\qquad \text{Square}$$

$$= \frac{25}{36}$$

Substitute $c = \frac{25}{36}$ into the trinomial and factor.

$$y^2 - \frac{5}{3}y + c \qquad\qquad \text{Substitute } c = \frac{25}{36}$$

$$y^2 - \frac{5}{3}y + \frac{25}{36} \qquad\qquad \text{Factor perfect square trinomial}$$

$$\left(y - \frac{5}{6}\right)^2 \qquad\qquad \text{Our Solution}$$

Check to verify factor is correct.

$$\left(y-\frac{5}{6}\right)^2 = \left(y-\frac{5}{6}\right)\left(y-\frac{5}{6}\right)$$

$$= y\left(y-\frac{5}{6}\right) - \frac{5}{6}\left(y-\frac{5}{6}\right)$$

$$= y^2 - \frac{5}{6}y - \frac{5}{6}y + \frac{25}{36}$$

$$= y^2 - \frac{10}{6}y + \frac{25}{36}$$

$$= y^2 - \frac{5}{3}y + \frac{25}{36} \quad \checkmark$$

Let us see how this problem can be done geometrically.

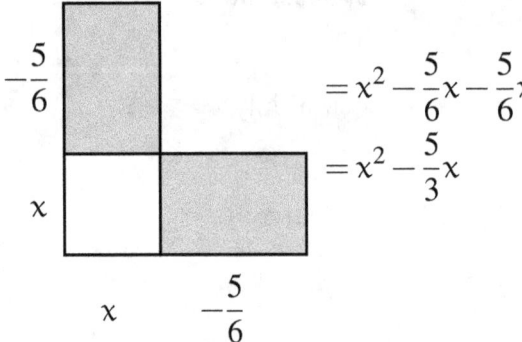

Cut the side with length "$-\frac{5}{3}$" units in half and connect the ends as follows.

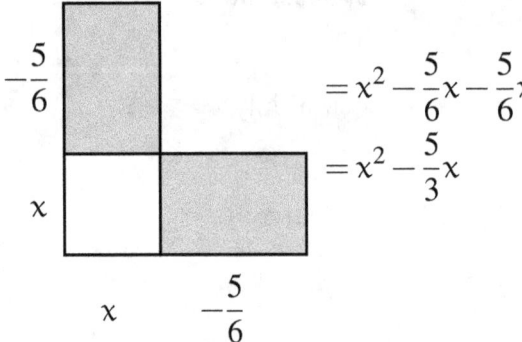

We now need to complete the square of the above figure by finding the area of the missing square in question.

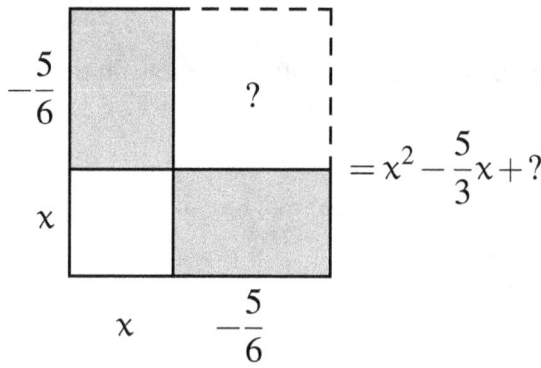

$$= x^2 - \frac{5}{3}x + ?$$

The missing square in question has sides of length $-\frac{5}{6}$. Its area is $\left(-\frac{5}{6}\right) \cdot \left(-\frac{5}{6}\right) = \frac{25}{36}$. Therefore, $\frac{25}{36}$ must be added to $y^2 - \frac{5}{3}y$ in order to form a perfect square trinomial and to complete the square of the figure.

Exercise 1 Class Example

Find the constant term, c, to make each expression into a perfect square trinomial. Then factor the trinomial.

a) $m^2 + 2m + c$

c) $p^2 - 5p + c$

b) $y^2 - 6y + c$

d) $n^2 + \frac{2}{3}n + c$

Exercise 2 You Try
Find the constant term, c, to make each expression into a perfect square trinomial. Then factor the trinomial.

a) $x^2 + 12x + c$ c) $m^2 + m + c$

b) $y^2 - 8y + c$ d) $n^2 - \dfrac{3}{5}n + c$

Solving Quadratic Equations by Completing the Square

Steps to Solving a Quadratic Equation of the form, $ax^2 + bx + c = 0$, by Completing the Square

1. Separate the constant term from the variable terms. Be sure variable terms are written in descending order.

2. If the coefficient of x^2, $a \neq 1$, divide each term by a. Otherwise, go to the next step.

3. Find the value, c, to complete the square, where $c = \left(\frac{1}{2}b\right)^2$.

4. Add the value of c to each side of the equation.

5. Factor the trinomial, on one side, and add the constant terms, on the other side. The trinomial should factor as a perfect square trinomial.

6. Solve for the variable by taking the square root of each side and using the square root property.

Example 6 Solve $x^2 + 6x = 16$ by completing the square. Be sure to simplify your answer. If the answer is irrational, round to the nearest tenth.

Solution.

Let us go through the steps in completing the square. Notice that $a = 1$, and $b = 6$.

1. Constant and variable terms are already separated $\qquad$ $x^2 + 6x \quad = \quad 16$

2. $a = 1$; move to the next step

3. Find the value of $c = \left(\frac{1}{2}b\right)^2$

   $c = \left(\frac{1}{2} \cdot 6\right)^2 = (3)^2 = 9$

4. Add $c = 9$ to each side of the equation $\qquad\qquad\qquad$ $x^2 + 6x + 9 \quad = \quad 16 + 9$

5. Factor the trinomial and add constant terms $\qquad\qquad$ $(x + 3)^2 \quad = \quad 25$

6. Take the square root of each side and simplify $\qquad$ $\sqrt{(x+3)^2} \quad = \quad \sqrt{25}$

   Solve the absolute value equation $\qquad\qquad\qquad\qquad$ $|x + 3| \quad = \quad 5$

   Solve for x $\qquad\qquad\qquad\qquad\qquad\qquad\qquad\qquad$ $x + 3 \quad = \quad \pm 5$

   $\qquad\qquad\qquad\qquad\qquad\qquad\qquad\qquad\qquad\qquad\quad$ $x \quad = \quad -3 \pm 5$

   Perform the indicated operation $\qquad\qquad\qquad$ $x = -3 + 5 \quad \text{or} \quad x = -3 - 5$

   Our Solution $\qquad\qquad\qquad\qquad\qquad\qquad\qquad\qquad$ $x = 2 \quad \text{or} \quad x = -8$

Verify that we have the correct solution by substituting our answer into the original quadratic equation, $x^2 + 6x = 16$.

When $x = 2$: $\qquad\qquad\qquad\qquad\qquad\qquad$ When $x = -8$:

$\quad (2)^2 + 6(2) \stackrel{?}{=} 16$ $\qquad\qquad\qquad\qquad$ $(-8)^2 + 6(-8) \stackrel{?}{=} 16$

$\qquad 4 + 12 = 16 \ \checkmark$ $\qquad\qquad\qquad\qquad\qquad$ $64 - 48 = 16 \ \checkmark$

Our solution to the quadratic equation is $x = 2$ or $x = -8$.

Note. The quadratic equation, $x^2 + 6x = 16$, can also be solved by factoring.

$\qquad\quad x^2 + 6x = 16$ $\qquad\qquad\qquad$ Move constant to other side of equation

$\quad x^2 + 6x - 16 = 0$ $\qquad\qquad\qquad$ Factor trinomial

$\quad (x + 8)(x - 2) = 0$ $\qquad\qquad\qquad$ Use zero-product property

$\qquad x + 8 = 0 \text{ or } x - 2 = 0$ $\qquad\qquad$ Solve for x

$\qquad\qquad x = -8 \text{ or } x = 2$ $\qquad\qquad$ Our Solution

From the above example, we see that solving the quadratic equation by factoring or completing the square yields the same answer but factoring is much simpler. However, not every quadratic equation can be solved by factoring. On the other hand, any quadratic equation can be solved by completing the square.

World View Note   Around 400 BC, the Babylonians developed an algorithmic approach to solving quadratic equations by completing the square. However, only positive solutions were considered. Their solution is geometric in nature. Solving quadratic equations this way focuses on the dimensions of a square. Hence, the term "quadratic" which means square.

Let us see how they solved $x^2 + 6x = 16$.

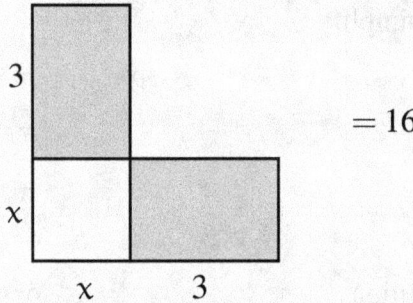

Cut the side with length, 6, in half and connect the ends as follows.

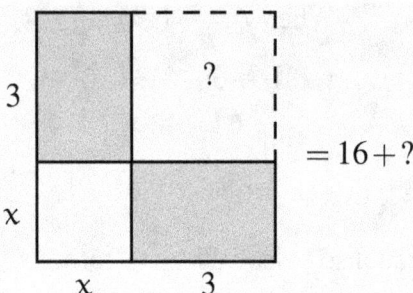

We now "complete the square" of the above figure by finding the area of the missing square in question.

The missing square in question has sides of length, 3. Therefore, its area is $3 \cdot 3 = 9$. The area, 9, must be added on the left figure to "complete the square." However, since we are working with an equation, if 9 is added on the left, 9 must also be added on the right as follows.

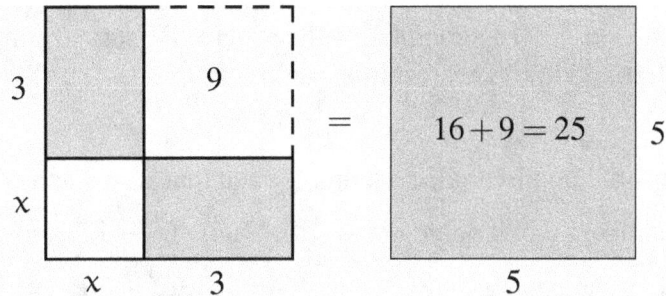

Both figures are equal to each other and both are squares. The figure on the left is a square with sides, $(x+3)$, and the figure on the right is also a square with length, 5. Equating the 2 sides, we get:

$$x+3=5$$
$$x=2$$

The positive solution to the quadratic equation $x^2 + 6x = 16$ is $x = 2$.

The next examples cannot be solved by factoring because the trinomials are not factorable. We will solve each equation by completing the square. Verification of solutions that are irrational will be done in the next chapter.

Example 7 Solve $x^2 - 8x - 2 = 0$ by completing the square. Be sure to simplify your answer. If the answer is irrational, round to the nearest hundredth.

Solution.
Let us go through the steps in completing the square. Notice that $a = 1$ and $b = -8$.

1. Separate the constant from the variable terms $\qquad x^2 - 8x - 2 \;=\; 0$

    Add 2 to each side of the equation $\qquad\qquad\qquad\quad x^2 - 8x \;=\; 2$

2. $a = 1$; move to the next step

3. Find the value of $c = \left(\frac{1}{2}b\right)^2$

    $c = \left(\frac{1}{2}\cdot -8\right)^2 = (-4)^2 = 16$

4. Add $c = 16$ to each side of the equation $\qquad x^2 - 8x + 16 \;=\; 2+16$

5. Factor the trinomial and add constant terms $\qquad (x-4)^2 \;=\; 18$

6. Take the square root of each side and simplify $\qquad \sqrt{(x-4)^2} \;=\; \sqrt{18}$

    Solve the absolute value equation $\qquad\qquad\qquad |x-4| \;=\; 3\sqrt{2}$

    Solve for x $\qquad\qquad\qquad\qquad\qquad\qquad\quad x-4 \;=\; \pm 3\sqrt{2}$

    Exact Solution $\qquad\qquad\qquad\qquad\qquad\qquad x \;=\; 4\pm 3\sqrt{2}$

    Approximate Solution $\qquad\qquad x \approx 8.24 \quad \text{or} \quad x \approx -0.24$

Example 8 Solve $3y^2 + 6y - 12 = 0$ by completing the square. Be sure to simplify your answer. If the answer is irrational, round to the nearest thousandth.

Solution.
Let us go through the steps in completing the square. Notice that $a = 3$ and $b = 6$.

1. Separate the constant from the variable terms $3y^2 + 6y - 12 \quad = \quad 0$

     Add 12 to each side of the equation $3y^2 + 6y \quad = \quad 12$

2. $a \neq 1$ so divide each term by $a = 3$ $y^2 + 2y \quad = \quad 4$

3. Find $c = \left(\frac{1}{2}b\right)^2$

     $c = \left(\frac{1}{2} \cdot 2\right)^2 = (1)^2 = 1$

4. Add $c = 1$ to each side of equation $y^2 + 2y + 1 \quad = \quad 4+1$

5. Factor the trinomial and add constant terms $(y + 1)^2 \quad = \quad 5$

6. Take the square root of each side and simplify $\sqrt{(y+1)^2} \quad = \quad \sqrt{5}$

     Solve the absolute value equation $|y + 1| \quad = \quad \sqrt{5}$

     Solve for y $y + 1 \quad = \quad \pm\sqrt{5}$

                  Exact Solution $y \quad = \quad -1 \pm \sqrt{5}$

             Approximate Solution $y \approx 1.236 \quad \text{or} \quad y \approx -3.236$

Exercise 3 Class Example
Solve each quadratic equation by completing the square. Be sure to simplify your answer. If the answer is irrational, round to the nearest thousandth.

a) $y^2 - 12y + 7 = 2$ b) $2g^2 + 16g + 4 = 0$

Exercise 4 **You Try**
Solve each quadratic equation by completing the square. Be sure to simplify your answer. If the answer is irrational, round to the nearest thousandth.

a) $x^2 - 6x - 7 = 0$ b) $5n^2 = 10n + 15$

Solving quadratic equations by completing the square will sometimes involve fractions. You will need to be comfortable with operations involving fractions. Let us take a look at a few examples.

Example 9 Solve $x^2 - 3x - 2 = 0$ by completing the square. Be sure to simplify your answer. If the answer is irrational, round to the nearest hundredth.

Solution.
Let us go through the steps in completing the square. Notice that $a = 1$ and $b = -3$.

1. Separate the constant from the variable terms $\qquad x^2 - 3x - 2 = 0$

    Add 2 to each side of equation $\qquad\qquad\qquad x^2 - 3x = 2$

2. $a = 1$; move to the next step

3. Find $c = \left(\frac{1}{2}b\right)^2$

    $c = \left(\frac{1}{2} \cdot -3\right)^2 = \left(-\frac{3}{2}\right)^2 = \frac{9}{4}$

4. Add $c = \frac{9}{4}$ to each side of equation $\qquad x^2 - 3x + \frac{9}{4} = 2 + \frac{9}{4}$

5. Factor the trinomial and add constant terms $\qquad \left(x - \frac{3}{2}\right)^2 = \frac{17}{4}$

6. Take the square root of each side and simplify $\qquad \sqrt{\left(x - \frac{3}{2}\right)^2} = \sqrt{\frac{17}{4}}$

    Solve the absolute value equation $\qquad\qquad \left|x - \frac{3}{2}\right| = \frac{\sqrt{17}}{2}$

    Solve for x $\qquad\qquad\qquad\qquad\qquad\qquad x - \frac{3}{2} = \pm\dfrac{\sqrt{17}}{2}$

    Exact Solution $\qquad\qquad\qquad\qquad\qquad\quad x = \frac{3}{2} \pm \frac{\sqrt{17}}{2}$

    or

    $$x = \frac{3 \pm \sqrt{17}}{2}$$

    Approximate Solution $\qquad x \approx 3.56 \quad \text{or} \quad x \approx -0.56$

Example 10 Solve $3y^2 = 5 - 2y$ by completing the square. Be sure to simplify your answer. If the answer is irrational, round to the nearest tenth.

Solution.
Let us go through the steps in completing the square.

1. Separate the constant from the variable terms $\qquad$ $3y^2 = 5-2y$

   Add 2y to each side of equation $\qquad$ $3y^2+2y = 5$

   Notice that $a=3$ and $b=2$

2. $a \neq 1$ so divide each term by $a=3$ $\qquad$ $y^2+\frac{2}{3}y = \frac{5}{3}$

3. Find $c = \left(\frac{1}{2}b\right)^2$

   $c = \left(\frac{1}{2}\cdot\frac{2}{3}\right)^2 = \left(\frac{1}{3}\right)^2 = \frac{1}{9}$

4. Add $c=\frac{1}{9}$ to each side of equation $\qquad$ $y^2+\frac{2}{3}y+\frac{1}{9} = \frac{5}{3}+\frac{1}{9}$

5. Factor the trinomial and add constant terms $\qquad$ $\left(y+\frac{1}{3}\right)^2 = \frac{16}{9}$

6. Take the square root of each side and simplify $\qquad$ $\sqrt{\left(y+\frac{1}{3}\right)^2} = \sqrt{\frac{16}{9}}$

   Solve the absolute value equation $\qquad$ $\left|y+\frac{1}{3}\right| = \frac{4}{3}$

   Solve for y $\qquad$ $y+\frac{1}{3} = \pm\frac{4}{3}$

   $\qquad$ $y = -\frac{1}{3}+\frac{4}{3}$ or $y=-\frac{1}{3}-\frac{4}{3}$

   Our Solution $\qquad$ $y=-1$ or $y=-\frac{5}{3}$

Example 11 Solve $2x^2 = 3x-7$ by completing the square. Be sure to simplify your answer. If the answer is irrational, round to the nearest thousandth.

Solution.
Let us go through each of the steps in completing the square.

1. Separate the constant from the variable terms $\qquad$ $2x^2 = 3x-7$

   Subtract 3x from each side of equation $\qquad$ $2x^2-3x = -7$

   Notice that $a=2$ and $b=-3$

2. $a \neq 1$ so divide each term by $a=2$ $\qquad$ $x^2-\frac{3}{2}y = -\frac{7}{2}$

3. Find $c = \left(\frac{1}{2}b\right)^2$

   $c = \left(\frac{1}{2}\cdot-\frac{3}{2}\right)^2 = \left(-\frac{3}{2}\right)^2 = \frac{9}{16}$

4. Add $c=\frac{9}{16}$ to each side of equation $\qquad$ $x^2-\frac{3}{2}x+\frac{9}{16} = -\frac{7}{2}+\frac{9}{16}$

5. Factor trinomial and add constant terms $\qquad$ $\left(x-\frac{3}{4}\right)^2 = -\frac{47}{4}$

6. Take square root of each side and simplify $\qquad$ $\sqrt{\left(x-\frac{3}{4}\right)^2} = \sqrt{-\frac{47}{4}}$

$\sqrt{-\frac{47}{4}}$ is not a real number! Therefore, the quadratic equation, $2x^2 = 3x - 7$ does not possess any real number solution. The solution is a complex number.

Exercise 5 Class Example

Solve the following quadratic equations by completing the square. Be sure to simplify your answer. If the answer is irrational, round to the nearest hundredth.

a) $x^2 + x - 5 = 0$ c) $2p^2 = p + 4$

b) $3n^2 + 9n = 12$ d) $y^2 = 8y - 20$

Exercise 6 **You Try**

Solve the following quadratic equations by completing the square. Be sure to simplify your answer. If the answer is irrational, round to the nearest hundredth.

a) $n^2 + 3n = 2$

c) $4y^2 = 24 - 20y$

b) $3p^2 - 6p + 3 = 0$

d) $x^2 + 4x - 4 = 12$

8.2: Exercises

Find the constant term c to make each expression into a perfect square trinomial. Then factor the trinomial.

1. $a^2 + 4a + c$

2. $x^2 - 12x + c$

3. $m^2 - 6m + c$

4. $n^2 + 10n + c$

5. $y^2 - y + c$

6. $x^2 + 5x + c$

7. $r^2 - \frac{1}{3}r + c$

8. $g^2 + \frac{4}{5}g + c$

Solve each quadratic equation by completing the square. Be sure to simplify answers. If answers are irrational, round to the nearest hundredth.

9. $n^2 - 8n + 7 = 0$

10. $y^2 + 4y = 12$

11. $8g^2 + 16g = 64$

12. $x^2 - 8x - 12 = 0$

13. $3x^2 - 6x + 48 = 0$

14. $8b^2 + 16b - 37 = 5$

15. $m^2 = -15 + 9m$

16. $v^2 = 14v + 36$

17. $5k^2 - 10k - 45 = 0$

18. $x^2 + 16x + 55 = 5$

19. $2k^2 - 4k - 10 = -2$

20. $b^2 + 7b = 3$

21. $4a^2 + 16a - 1 = 0$

22. $5y^2 - 8y - 4 = 1$

23. $3p^2 = 2p + 6$

24. $3w^2 = 2 - w$

8.3   The Quadratic Formula

Objective: To solve quadratic equations using the quadratic formula

Sometimes solving a quadratic equation by factoring or completing the square can be nearly impossible or cumbersome, especially when the leading coefficient of the quadratic expression is not 1. In those cases, it may be convenient to use a shortcut, called *the quadratic formula*.

Quadratic Formula Derivation

The quadratic formula is derived by completing the square of a quadratic equation in the form $ax^2 + bx + c = 0$. The steps may look overwhelming but they are exactly the steps used in completing the square except for the trinomial coefficients being non-numeric.

1. Separate the constant from the variable terms

$$ax^2 + bx + c = 0$$

Subtract c from each side of equation

$$ax^2 + bx = -c$$

2. $a \neq 1$ so divide each term by a

$$x^2 + \frac{b}{a}x = -\frac{c}{a}$$

3. Find $\left(\frac{1}{2}\cdot\frac{b}{a}\right)^2 = \left(\frac{b}{2a}\right)^2 = \frac{b^2}{4a^2}$

4. Add $\frac{b^2}{4a^2}$ to each side of equation

$$x^2 + \frac{b}{a}x + \frac{b^2}{4a^2} = \frac{b^2}{4a^2} - \frac{c}{a}$$

5. Factor trinomial and find LCD for fractions

$$\left(x+\frac{b}{2a}\right)^2 = \frac{b^2}{4a^2} - \frac{4ac}{4a^2}$$

Combine fractions

$$\left(x+\frac{b}{2a}\right)^2 = \frac{b^2-4ac}{4a^2}$$

6. Take the square root of each side and simplify

$$\sqrt{\left(x+\frac{b}{2a}\right)^2} = \sqrt{\frac{b^2-4ac}{4a^2}}$$

Solve the absolute value equation

$$\left|x+\frac{b}{2a}\right| = \frac{\sqrt{b^2-4ac}}{\sqrt{4a^2}}$$

Solve for x

$$x+\frac{b}{2a} = \pm\frac{\sqrt{b^2-4ac}}{2a}$$

Quadratic Formula written as separate fractions

$$x = -\frac{b}{2a} \pm \frac{\sqrt{b^2-4ac}}{2a}$$

Quadratic Formula written as a single fraction

$$x = \frac{-b \pm \sqrt{b^2-4ac}}{2a}$$

This formula is important because we can use it to solve any quadratic equation: oce the quadratic equation is in standard form, $ax^2 + bx + c = 0$, we identify a, b, and c, and then substitute those values into the quadratic formula to get our solution, $x = \frac{-b\pm\sqrt{b^2-4ac}}{2a}$.

The Quadratic Formula

$$\text{If } ax^2 + bx + c = 0 \text{, then } x = \frac{-b \pm \sqrt{b^2 - 4ac}}{2a}$$

Steps for solving a quadratic equation using the quadratic formula:

1. Write the quadratic equation in standard form, $ax^2 + bx + c = 0$.

2. Identify the values of a (the coefficient of x^2), b (the coefficient of x) and c (the constant term).

3. Put the values of a, b, and c into the quadratic formula, $x = \dfrac{-b \pm \sqrt{b^2 - 4ac}}{2a}$.

4. Use the order of operations to simplify the solutions.

Example 1 Solve $x^2 + 3x + 2 = 0$ using the quadratic formula. Be sure to simplify your answer. If the answer is irrational, round to the nearest hundredth.

Solution.
Let us go through each of the steps in solving the quadratic equation by using the quadratic formula.

1.	Equation already in standard form	$x^2 + 3x + 2 = 0$
2.	Identify the values of a, b, and c	$a = 1$, $b = 3$, $c = 2$
3.	Put the values into quadratic formula	$x = \dfrac{-(3) \pm \sqrt{(3)^2 - 4(1)(2)}}{2(1)}$
4.	Simplify the solution	$x = \dfrac{-3 \pm \sqrt{9 - 8}}{2}$
		$x = \dfrac{-3 \pm \sqrt{1}}{2}$
	Rational answers	$x = \dfrac{-3 \pm 1}{2}$
	Perform indicated operation	$x = \dfrac{-3 + 1}{2}$ or $x = \dfrac{-3 - 1}{2}$
		$x = \dfrac{-2}{2}$ or $x = \dfrac{-4}{2}$
	Our Solution	$x = -1$ or $x = -2$

Verify that we have the correct solution by substituting our answer into the original quadratic

equation, $x^2 + 3x + 2 = 0$.

When $x = -1$:

$(-1)^2 + 3(-1) + 2 \stackrel{?}{=} 0$

$1 - 3 + 2 = 0$ ✓

When $x = -2$:

$(-2)^2 + 3(-2) + 2 \stackrel{?}{=} 0$

$4 - 6 + 2 = 0$ ✓

Our solution to the quadratic equation is $x = -1$ or $x = -2$.

Note. The above example can also be solved by factoring or completing the square. All of them will yield the same answer. Verify it.

Example 2 Solve $2y^2 = 2y + 1$ using the quadratic formula. Be sure to simplify your answer. If the answer is irrational, round to the nearest hundredth.

Solution.

Let us go through each of the steps in solving the quadratic equation by using the quadratic formula.

1. Put the equation in standard form $\qquad 2y^2 = 2y + 1$

    Subtract 2y and 1 from each side $\qquad 2y^2 - 2y - 1 = 0$

2. Identify the values of a, b,and c $\qquad a = 2, b = -2, c = -1$

3. Put the values into the quadratic formula $\quad y = \dfrac{-(-2) \pm \sqrt{(-2)^2 - 4(2)(-1)}}{2(2)}$

3. Put the values into quadratic formula $\quad m = \dfrac{-(0) \pm \sqrt{(0)^2 - 4(3)(2)}}{2(3)}$

4. Simplify the solution $\qquad y = \dfrac{2 \pm \sqrt{4 + 8}}{4}$

    $\qquad y = \dfrac{2 \pm \sqrt{12}}{4}$

    Factor 2 from numerator $\qquad y = \dfrac{2 \pm 2\sqrt{3}}{4}$

    Simplify fraction $\qquad y = \dfrac{2(1 \pm \sqrt{3})}{4}$

    Exact Solution $\qquad y = \dfrac{1 \pm \sqrt{3}}{2}$

    Approximate Solution $\qquad y \approx 1.37$ or $y \approx -0.37$

Example 3 Solve $3m^2 = -2$ using the quadratic formula. Be sure to simplify your answer. If the answer is irrational, round to the nearest hundredth.

Solution.

Let us go through each of the steps in solving the quadratic equation by using the quadratic formula.

1. Put the equation in standard form $3m^2 = -2$

    Add 2 to each side $3m^2 + 2 = 0$

2. Identify the values of a, b, and c $a = 3, b = 0, c = 2$

3. Put the values into quadratic formula $m = \dfrac{-(0) \pm \sqrt{(0)^2 - 4(3)(2)}}{2(3)}$

4. Simplify the solution $m = \dfrac{0 \pm \sqrt{0 - 24}}{6}$

                                               $m = \dfrac{\pm\sqrt{-24}}{6}$

$\dfrac{\pm\sqrt{-24}}{6}$ is not a real number! Therefore, the quadratic equation, $3m^2 = -2$ does not possess any real number solution. The solution is a complex number.

Example 4 Solve $12n - 4 = 9n^2$ using the quadratic formula. Be sure to simplify your answer. If the answer is irrational, round to the nearest hundredth.

Solution.

Let us go through each of the steps in solving the quadratic equation by using the quadratic formula.

1. Put the equation in standard form $12n - 4 = 9n^2$

    Subtract 12n and add 4 to each side $0 = 9n^2 - 12n + 4$

2. Identify the values of a, b, and c $a = 9, b = -12, c = 4$

3. Put the values into the quadratic formula $n = \dfrac{-(-12) \pm \sqrt{(-12)^2 - 4(9)(4)}}{2(9)}$

4. Simplify the solution $n = \dfrac{12 \pm \sqrt{144 - 144}}{18}$

                                                $n = \dfrac{12 \pm \sqrt{0}}{18}$

    Simplify fraction $n = \dfrac{12 \pm 0}{18}$

    Solution $n = \dfrac{2}{3}$

Note. In solving quadratic equations, sometimes we get 2 distinct solutions, sometimes no real solutions and sometimes one unique solution, as seen in the above example.

Exercise 1 **Class Example**

Solve the following quadratic equations using the quadratic formula. Be sure to simplify your answer. If the answer is irrational, round to the nearest hundredth.

a) $x^2 + 6x + 8 = 0$

c) $\frac{2}{3}y^2 = \frac{4}{9}y + \frac{1}{3}$

b) $3n^2 - 4n = 5$

d) $0.5m^2 = 0.3$

Exercise 2 **You Try**

Solve the following quadratic equations using the quadratic formula. Be sure to simplify your answer. If the answer is irrational, round to the nearest hundredth.

a) $x^2 - 3x - 5 = 0$

c) $5h - 3 = 6h^2$

b) $2n^2 = 7$

d) $\frac{2}{5}y^2 + \frac{2}{5}y + \frac{1}{10}$

World View Note This alternate derivation of the quadratic formula was known to the Hindus as early as AD 1025. It was still being taught as late as 1905 but was somehow replaced by the derivation done at the beginning of this section. This alternate derivation involves completing the square but does not require the leading coefficient to be 1. You will find that it is shorter, has simpler computations and does not involve fractions until the last step.

Put the equation in standard form	$ax^2 + bx + c$	$= 0$
Multiply each term by $4a$	$4a^2x^2 + 4abx + 4ac$	$= 0$
Separate variable and constant terms	$4ax^2 + 4bx$	$= -4ac$
Add b^2 to each side of equation	$4ax^2 + 4abx + b^2$	$= b^2 - 4ac$
Factor perfect square trinomial	$(2ax + b)^2$	$= b^2 - 4ac$
Take the square root of each side	$\sqrt{(2ax + b)^2}$	$= \sqrt{b^2 - 4ac}$
Solve the absolute value equation	$|2ax + b|$	$= \sqrt{b^2 - 4ac}$
Subtract b from each side	$2ax + b$	$= \pm\sqrt{b^2 - 4ac}$
Divide each side by $2a$	$2ax$	$= -b \pm \sqrt{b^2 - 4ac}$
We have the Quadratic Formula!	x	$= \dfrac{-b \pm \sqrt{b^2 - 4ac}}{2a}$

8.3: Exercises

Solve each equation using the quadratic formula and simplify your answers. If answers are irrational, round to the nearest hundredth.

1. $y^2 + 3y - 1 = 0$

2. $v^2 - 4v - 5 = -8$

3. $3p^2 - 5p + 3 = 4p^2$

4. $1 - \dfrac{2}{3}m^2 = 0$

5. $12g = 9g^2 + 4$

6. $y^2 - 4y = 1$

7. $3r^2 = 2r + 1$

8. $2x + 15 = 2x^2$

9. $m^2 - 14m + 55 = 0$

10. $k^2 = 3k + 5$

11. $2n^2 + 7n = 49$

12. $\dfrac{2}{5}b^2 = \dfrac{3}{5} - b$

13. $0.3n^2 - 1 = 0$

14. $r^2 + 4 = -6r$

15. $y^2 + \dfrac{1}{4} = -y$

16. $3v^2 = 2 + 3v$

17. $0 = 2x^2 + 5x + 3$

18. $1.6n^2 + 2.4n + 0.9 = 0$

8.4   Strategies for solving Quadratic Equations

Objective: To develop strategies of solving quadratic equations efficiently

We have seen three approaches to solving a quadratic equation.

1. By Factoring (if possible)

2. Using the Square Root Property or Completing the Square

3. Using the Quadratic Formula

Each technique has its advantages and limitations. In this section, we seek to develop strategies to determine which method will lead to the most efficient solution. Let us solve a few equations using the three methods and then summarize what we find.

Example 1 Solve the quadratic equation, $x^2 + 2x - 35 = 0$ using the three different techniques. Give answer in exact form. Then fill the table with pros and cons of each method and discuss your preference.

Solution.

1. **By Factoring (if possible)**

$x^2 + 2x - 35 = 0$	Factor trinomial
$(x+7)(x-5) = 0$	Use Zero-Factor Property
$x+7 = 0$ or $x-5 = 0$	Solve for x
$x = -7$ or $x = 0$	Our Solution

2. **Using the Square Root Property or Completing the Square**

$x^2 + 2x - 35 = 0$	Separate the constant from variable terms
$x^2 + 2x = 35$	Notice that $a = 1$ and $b = 2$
	Add $\left(\frac{1}{2}b\right)^2 = \left(\frac{1}{2} \cdot 2\right)^2 = (1)^2 = 1$ to each side
$x^2 + 2x + 1 = 35 + 1$	Factor trinomial and add constant terms
$(x+1)^2 = 36$	Take the square root of each side
$\sqrt{(x+1)^2} = \sqrt{36}$	Simplify
$|x+1| = 6$	Solve the absolute value equation
$x+1 = \pm 6$	Subtract 1 from each side
$x = -1 \pm 6$	Rational answer; Simplify solution
$x = -1+6$ or $x = -1-6$	Perform indicated operation
$x = 5$ or $x = -7$	Our Solution

3. **Using the Quadratic Formula**

$$x^2 + 2x - 35 = 0$$ Equation in standard form

$$x = \frac{-b \pm \sqrt{b^2 - 4ac}}{2a}$$ Quadratic Formula

$$x = \frac{-(2) \pm \sqrt{(2)^2 - 4(1)(-35)}}{2(1)}$$ Substitute $a = 1, b = 2, c = -35$

$$x = \frac{-2 \pm \sqrt{4 + 140}}{2}$$ Add radicand

$$x = \frac{-2 \pm \sqrt{144}}{2}$$ Simplify radical

$$x = \frac{-2 \pm 12}{2}$$ Rational answer

$$x = \frac{-2 + 12}{2} \text{ or } x = \frac{-2 - 12}{2}$$ Simplify solution

$$x = \frac{10}{2} \text{ or } x = \frac{-14}{2}$$ Simplify fraction

$$x = 5 \text{ or } x = -7$$ Our Solution

Technique	Pros	Cons
Factoring	leads directly to solution with minimal steps	
Square Root Property or Completing the Square		quadratic equation has to be rewritten to complete the square and there are a number of simplifying steps
Quadratic Formula		there are a number of simplifying steps

Example 2 Solve the quadratic equation, $(x-5)^2 - 4 = 3$ using the three different techniques. Give the answer in exact form. Then fill the table with pros and cons of each method and discuss your preference.

Solution.

1. **By Factoring (if possible)**
   Rewrite the quadratic equation so it is in standard form.

   $$(x-5)^2 - 4 = 3 \qquad \text{Expand } (x-5)^2$$
   $$x^2 - 10x + 25 - 4 = 3 \qquad \text{Combine like terms}$$
   $$x^2 - 10x + 21 = 3 \qquad \text{Subtract 3 from each side}$$
   $$x^2 - 10x + 18 = 0 \qquad \text{Factor trinomial}$$

   The trinomial is not factorable. We cannot find two integers whose product is 18 and whose sum is -10. Thus, we cannot solve this equation by factoring.

2. **Using the Square Root Property or Completing the Square**
   There is no need to expand the binomial term to complete the square.

   $$(x-5)^2 - 4 = 3 \qquad \text{Isolate the square term by adding 4 to each side}$$
   $$(x-5)^2 = 7 \qquad \text{Take the square root of each side}$$
   $$\sqrt{(x-5)^2} = \sqrt{7} \qquad \text{Simplify}$$
   $$|x-5| = \sqrt{7} \qquad \text{Solve the absolute value equation}$$
   $$x - 5 = \pm\sqrt{7} \qquad \text{Add 5 to each side}$$
   $$x = 5 \pm \sqrt{7} \qquad \text{Our Solution}$$

3. **Using the Quadratic Formula**
   We need to rewrite the quadratic equation in standard form by first expanding $(x-5)^2$. This was already done in the factoring technique above.

$$x^2 - 10x + 18 = 0 \qquad\qquad\qquad \text{Equation in standard form}$$

$$x = \frac{-b \pm \sqrt{b^2 - 4ac}}{2a} \qquad\qquad \text{Quadratic Formula}$$

$$x = \frac{-(-10) \pm \sqrt{(-10)^2 - 4(1)(18)}}{2(1)} \qquad \text{Substitute } a = 1, b = -10, c = 18$$

$$x = \frac{10 \pm \sqrt{100 - 72}}{2} \qquad\qquad \text{Subtract radicand}$$

$$x = \frac{10 \pm \sqrt{28}}{2} \qquad\qquad\qquad \text{Simplify radical}$$

$$x = \frac{10 \pm 2\sqrt{7}}{2} \qquad\qquad\qquad \text{Factor GCF from numerator}$$

$$x = \frac{2(5 \pm 2\sqrt{7})}{2} \qquad\qquad\qquad \text{Simplify fraction}$$

$$x = 5 \pm \sqrt{7} \qquad\qquad\qquad\qquad \text{Our Solution}$$

Technique	Pros	Cons
Factoring		trinomial is not factorable so technique cannot be used
Square Root Property or Completing the Square	most efficient way of finding the solution because square term can easily be isolated and removed using the square root property	
Quadratic Formula		have to rewrite equation in standard form first and there are a number of simplifying steps

Exercise 1 Class Example

Solve the following quadratic equation using the three different techniques. Give answer in exact form. Then fill the table with pros and cons of each method and discuss your preference.

a) $x^2 - 6x = -5$

c) $\dfrac{1}{2}(x+7)^2 = 6$

b) $x^2 - 6x + 4 = 0$

d) $4x^2 - x = 1$

Exercise 2 You Try

Solve the following quadratic equation using the three different techniques. Give answer in exact form. Then fill the table with pros and cons of each method and discuss your preference.

a) $x^2 = 3x$

c) $9x^2 = 32$

b) $x^2 = 3x + 1$

d) $x^2 - 2x = 5$

SUMMARY:

Now that we have worked through a number of examples, there are a few key observations we should keep in mind when solving a quadratic equation.

1. Is there a **single square term** in the equation, such as, $x^2 = 3$, $2x^2 = 9$, or $(3x+4)^2 + 2 = 8$? If so, then the **square root property** will typically lead to the most efficient solution.

2. Are there **both a square term and a linear term** in the equation, such as, $x^2 = 3x$, $2x^2 + 5x = 9$, or $(3x+4)^2 + 2x = 8$? If so, then you'll want to do the following.

   (a) Write the equation in standard form and try to solve the equation by **Factoring**.

   (b) If the trinomial is not factorable, use either the **Quadratic Formula** or **Complete the Square**. Note that if the coefficient of the square term is not 1 and/or the resulting linear term is odd, then completing the square will be quite messy. So the quadratic formula would be the ideal method to use.

Exercise 3 **Class Example**

Identify the most efficient technique to solve the given quadratic equation and indicate the main steps you would take in solving the equation. **Do not solve** the quadratic equations.

a) $4x^2 = 3x + 10$

c) $2p^2 = p + 5$

b) $7 - y^2 = 3$

d) $w^2 + 4w = 4$

Exercise 4 You Try

Identify the most efficient technique to solve the given quadratic equation and indicate the main steps you would take in solving the equation. **Do not solve** the quadratic equations.

a) $3p - 5p^2 = -4$

c) $c^2 - 8c = 4$

b) $2y^2 = 35 + 9y$

d) $2w^2 + 3w = 0$

8.4: Exercises

1. Solve each equation using the indicated technique. Then state which technique you find most efficient.

    i. Factoring (if possible)

    ii. Square Root Property / Completing the Square

    iii. Quadratic Formula

(a) $x^2 - 6x + 8 = 0$ (d) $z^2 + 4z = 3$

(b) $y^2 = 12y$ (e) $(x-6)^2 - 3 = 9$

(c) $9b^2 - 81 = 0$ (f) $2v^2 = 12 - 5v$

2. Identify the most efficient technique to solve the given quadratic equation and indicate the main steps you would take in practice to solve the equation. You don't have to solve the equation for this problem (though you may want to, for extra practice).

(a) $3y^2 - 8y - 35 = 0$ (e) $A^2 + 10A = 2$

(b) $x(x+6) = 16$

(c) $3z^2 = 12$ (f) $5 + (2b-7)^2 = 10$

(d) $3z^2 = 12z$ (g) $5 + (2b-7)^2 = 10b$

3. Solve each equation using any method of your choice.

(a) $x(x-3) = 0$ (k) $y(3y-2) = 8$

(b) $x(x-3) = 4$ (l) $x^2 + 6x = 9$

(c) $3y^2 = 9y$ (m) $-6B^2 - 12B + 50 = 4B - 20$

(d) $3y^2 = 9$

(e) $4x^2 + 4x - 3 = 0$ (n) $-2x(2-x) = 8 + 4x$

(f) $(2x-7)^2 = 5$ (o) $3w^2 - 4w + 11 = 0$

(g) $(2x-7)^2 = 0$ (p) $\frac{1}{2}t^2 + 2t = 6$

(h) $6v^2 - v = 12$

(i) $4b^2 - 8b - 7 = 0$ (q) $200x^2 - 300x = 400$

(j) $9c^2 + 4 = 12c$ (r) $\frac{3}{4}(b-5)^2 = 6$

8.5   Graphs of Quadratic Equations

Objective: To graph quadratic equations

In this section, we will learn how to graph quadratic equations, $y = ax^2 + bx + c$. The graph of a quadratic equation consists of all sets of (x, y) pairs that make the quadratic equation true. For example, if we consider the quadratic equation, $y = 3x^2 - 2x + 1$.

- the point $(2, 9)$ is on the graph of $y = 3x^2 - 2x + 1$, since $2 = 3(2)^2 - 2(2) + 1$.

- the point $(1, 1)$ is not on the graph of $y = 3x^2 - 2x + 1$ since $1 \neq 3(1)^2 - 2(1) + 1$.

Example 1 Graph the equation, $y = x^2 + 1$ by first filling in the table. Then use the ordered pairs resulting from the computation to sketch the graph.

x	$y = x^2 + 1$	Ordered Pair
-2		
-1		
0		
1		
2		

Solution.

Let us start by filling in the table of solutions for this equation.

x	$y = x^2 + 1$	Ordered Pair
-2	$y = (-2)^2 + 1 = 5$	$(-2, 5)$
-1	$y = (-1)^2 + 1 = 2$	$(-1, 2)$
0	$y = (0)^2 + 1 = 1$	$(0, 1)$
1	$y = (1)^2 + 1 = 2$	$(1, 2)$
2	$y = (2)^2 + 1 = 5$	$(2, 5)$

We obtain our graph by plotting the ordered pairs we found and connecting them with a smooth curve.

The graph of a quadratic equation will not be a line as we can see from the points we obtained.

This shape is called a parabola.

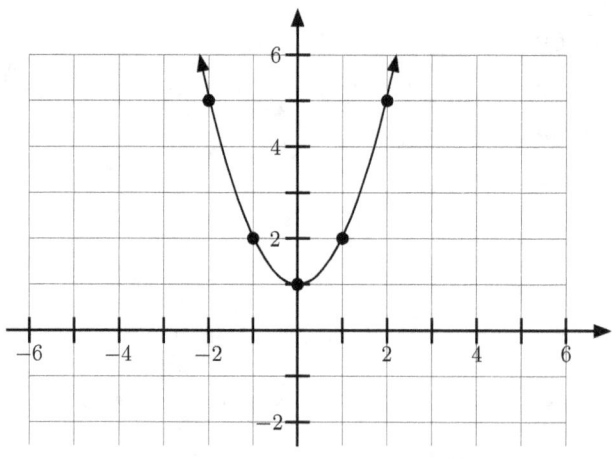

Exercise 1 **Class Example**

Graph the equation, $y = 2x^2 - 1$ by first filling in the table. Then use the ordered pairs resulting from the computation to sketch the graph.

x	$y = 2x^2 - 1$	**Ordered Pair**
-2		
-1		
0		
1		
2		

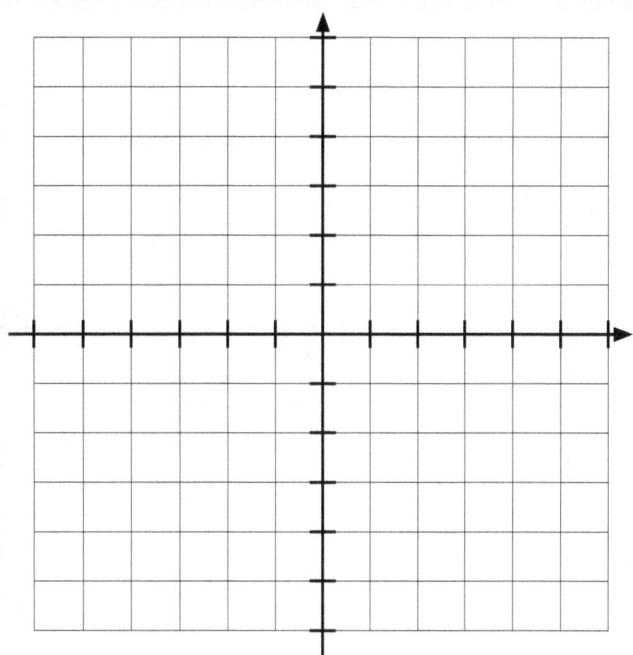

Exercise 2 **You Try**

Graph the equation, $y = -x^2$ by first filling in the table. Then use the ordered pairs resulting from the computation to sketch the graph.

x	$y = -x^2$	**Ordered Pair**
-2		
-1		
0		
1		
2		

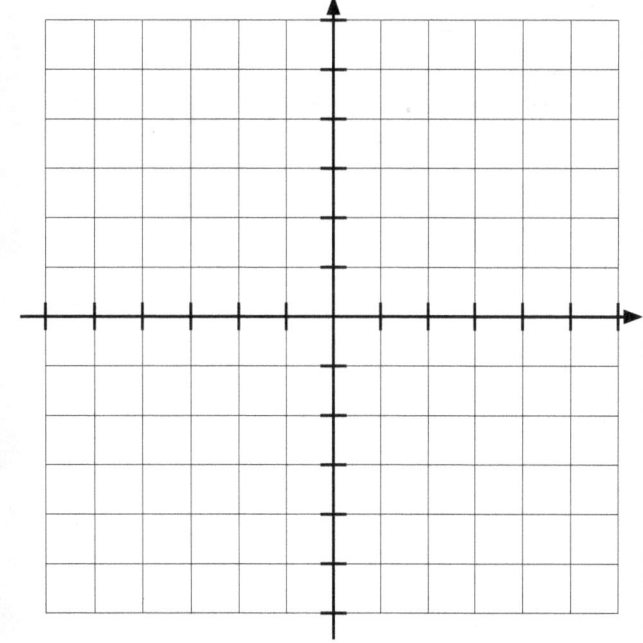

Exercise 3 **Class Example**

Graph the equation, $y = -x^2 + 2x + 3$ by first filling in the table. Then use the ordered pairs resulting from the computation to sketch the graph.

x	$y = -x^2 + 2x + 3$	Ordered Pair
-1		
0		
1		
2		
3		

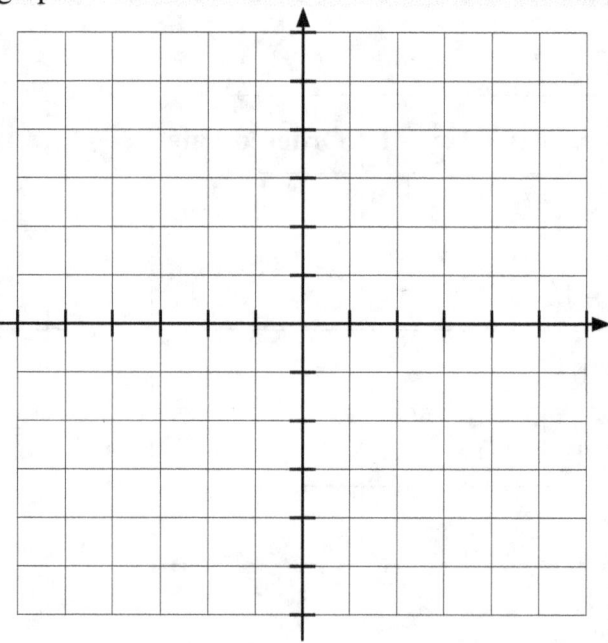

Exercise 4 **You Try**

Graph the equation, $y = x^2 - 4x + 4$ by first filling in the table. Then use the ordered pairs resulting from the computation to sketch the graph.

x	$y = x^2 - 4x + 4$	Ordered Pair
-1		
0		
2		
4		
5		

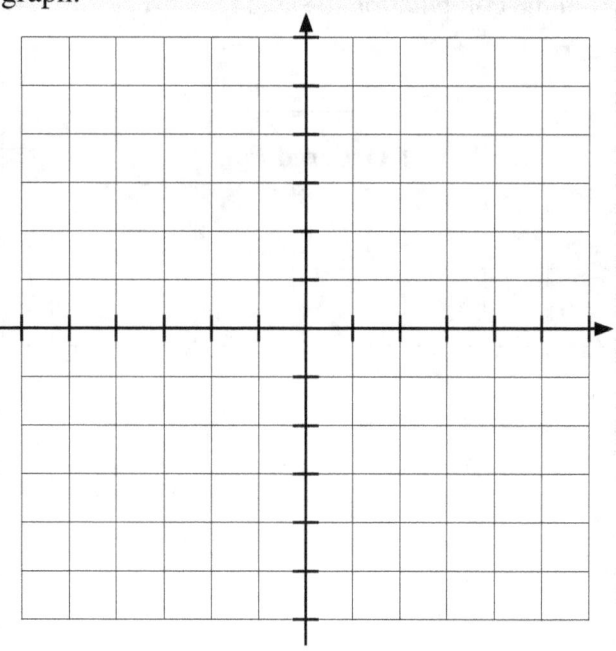

Properties of the Graph of a Quadratic Equation

Notice that the graph of a quadratic equation, $y = ax^2 + bx + c$, is not a line. It is a curve called a parabola that has the following properties.

- The graph of the parabola either opens up or opens down, depending on the sign of the leading coefficient, a, of the quadratic term.

- The parabola has a highest point, if it opens down, or a lowest point, if it opens up. This point is called the **vertex** of the parabola.

- The parabola always has a y-intercept.

- The parabola can have 2 x-intercepts, 1 x-intercept, or no x-intercept.

- The parabola is symmetric about a vertical line that goes through the vertex. This line is called the line of symmetry.

- If the parabola has 2 x-intercepts, then the x-coordinate of the vertex is at the midpoint between the two x-intercepts.

- If the parabola has 1 x-intercept, then the vertex is the same as the x-intercept.

Vertex and Line of Symmetry

The vertex of a parabola is the highest or lowest point of the parabola. If the parabola has two x-intercepts, then, as noticed above, the x-coordinate of the vertex is the midpoint between them. Since the quadratic formula gives us the values of the x-intercepts, and can be written as

$$x = \frac{-b}{2a} \pm \frac{\sqrt{b^2 - 4ac}}{2a}$$

we can see that the midpoint between them is precisely $x = \dfrac{-b}{2a}$.

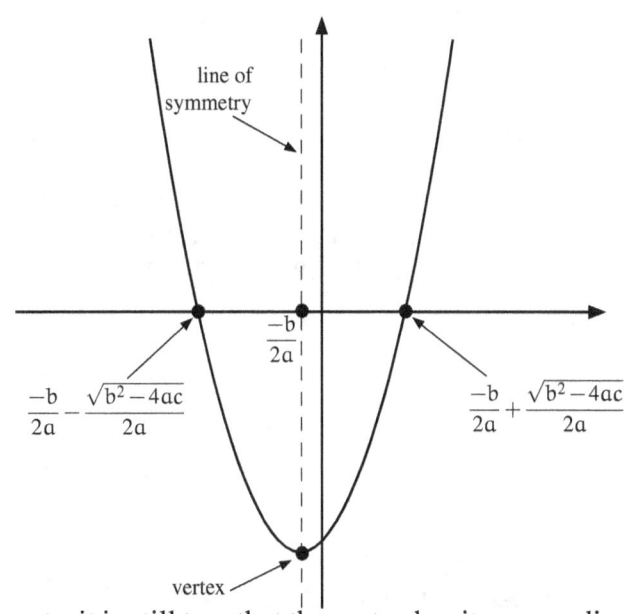

Even if the parabola does not have two x-intercepts, it is still true that the vertex has its x-coordinate at $x = \frac{-b}{2a}$. To find the y-coordinate of the vertex, substitute the x-value into the quadratic equation.

The line of symmetry is the vertical line that goes through the vertex. It's equation is $x = -\frac{b}{2a}$. For any point on the parabola to the left (or right) of the line of symmetry, there is another point with the same distance from the line of symmetry, but to the right (or left) of the line of symmetry.

From now on, we will focus on finding characteristics of a parabola, instead of finding a list of points on the graph. We will start with parabolas with 2 distinct x-intercepts.

Example 2 Sketch the graph of the parabola, $y = x^2 + 2x - 8$. Be sure to do the following and write all points as ordered pairs.

a) Determine if the parabola opens upward or downward.

b) Find the y-intercept.

c) Find any x-intercepts. If the answer is irrational, round to the nearest hundredth.

d) Find the vertex.

e) Find the line of symmetry.

f) Find the point symmetric to the y-intercept.

Solution.

First identify the coefficients $a = 1$, $b = 2$ and $c = -8$.

a) The leading coefficient, $a = 1 > 0$. Therefore, the parabola opens upward.

b) To find the y-intercept, let $x = 0$.
$$y = (0)^2 + 2(0) - 8 = -8$$
So, the y-intercept is $(0, -8)$.

c) To find x-intercepts, let $y = 0$ and solve for x. In this case, we can solve for x by factoring.

$$x^2 + 2x - 8 = 0$$
$$(x - 2)(x + 4) = 0$$
$$x - 2 = 0 \text{ or } x + 4 = 0$$
$$x = 2 \text{ or } x = -4$$

The x-intercepts are $(2, 0)$ and $(-4, 0)$.

d) The x-coordinate of the vertex is $x = -\dfrac{b}{2a} = -\dfrac{2}{2(1)} = -1$

Substitute $x = -1$ into the quadratic equation to find the y-coordinate of the vertex.
$$y = (-1)^2 + 2(-1) - 8 = 1 - 2 - 8 = -9$$

The vertex is $(-1, -9)$.

e) The line of symmetry is $x = -1$. It is a vertical line going through the vertex.

f) The y-intercept, $(0, -8)$, is 1 unit to the right of the line of symmetry, $x = -1$. By symmetry, there is a point 1 unit to the left of $x = -1$. That point is $(-2, -8)$ and is symmetric to the y-intercept.

We can now use all this information to graph the parabola.

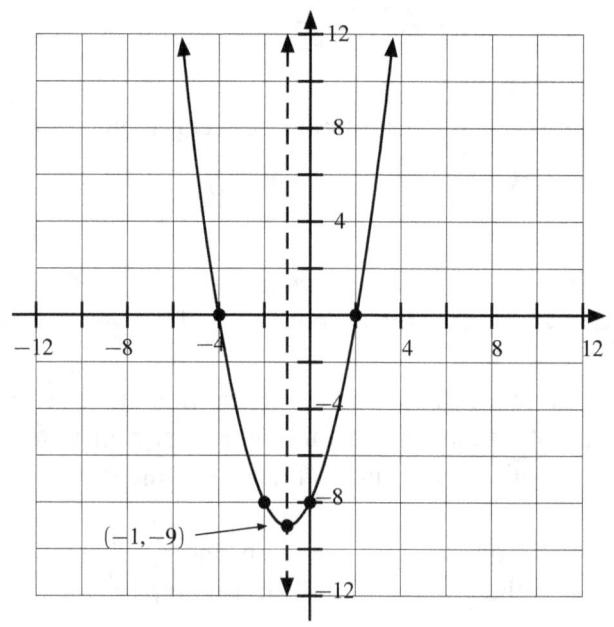

$(-1, -9)$

Example 3 Sketch the graph of the parabola, $y = -x^2 + 2x + 3$. Be sure to do the following and write all points as ordered pairs.

a) Determine if the parabola opens upward or downward.

b) Find the y-intercept.

c) Find any x-intercepts. If the answer is irrational, round to the nearest hundredth.

d) Find the vertex.

e) Find the line of symmetry.

f) Find the point symmetric to the y-intercept.

Solution.

First identify the coefficients $a = -1$, $b = 2$ and $c = 3$.

a) The leading coefficient, $a = -1 < 0$. Therefore, the parabola opens downward.

b) To find the y-intercept, let $x = 0$.
$y = -(0)^2 + 2(0) + 3 = 3$
So, the y-intercept is $(0,3)$.

c) To find x-intercepts, let $y = 0$ and solve for x. In this case, we can solve for x by factoring.

$$-x^2 + 2x + 3 = 0 \qquad \text{Multiply each term by } -1$$
$$x^2 - 2x - 3 = 0 \qquad \text{Factor the trinomial}$$
$$(x - 3)(x + 1) = 0 \qquad \text{Use the zero-product property}$$
$$x - 3 = 0 \text{ or } x + 1 = 0 \qquad \text{Solve for x}$$
$$x = 3 \text{ or } x = -1 \qquad \text{Our Solution}$$

The x-intercepts are $(3,0)$ and $(-1,0)$.

d) The x-coordinate of the vertex is $x = -\dfrac{b}{2a} = -\dfrac{2}{2(-1)} = 1$
Substitute $x = 1$ into the quadratic equation to find the y-coordinate of the vertex.
$y = -(1)^2 + 2(1) + 3 = -1 + 2 + 3 = 4$
The vertex is $(1,4)$.

e) The line of symmetry is $x = 1$. It is a vertical line going through the vertex.

f) The y-intercept, $(0,3)$, is 1 unit to the left of the line of symmetry, $x = 1$. By symmetry, there is a point 1 unit to the right of $x = 1$. That point is $(2,3)$ and is symmetric to the y-intercept.

We can now use all this information to graph the parabola.

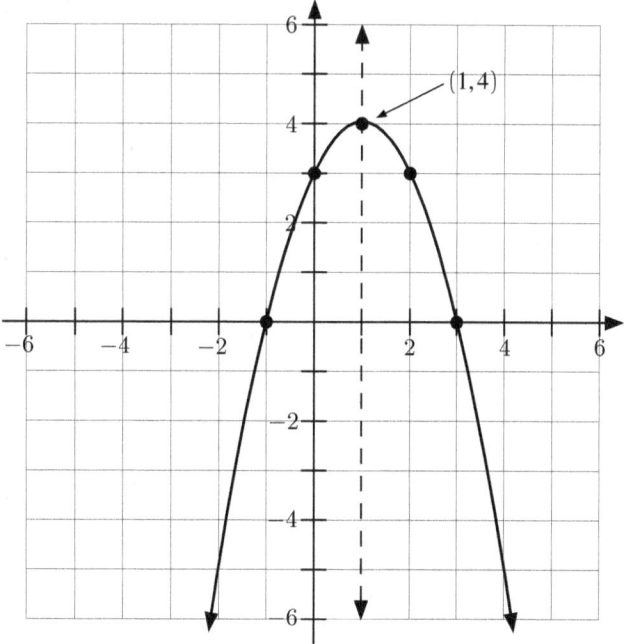

Exercise 5 Class Example

Sketch the graph of the parabola, $y = x^2 + 4x - 5$. Be sure to do the following and write all points as ordered pairs.

a) Determine if the parabola opens upward or downward.

b) Find the y-intercept.

c) Find any x-intercepts. If the answer is irrational, round to the nearest hundredth.

d) Find the vertex.

e) Find the line of symmetry.

f) Find the point symmetric to the y-intercept.

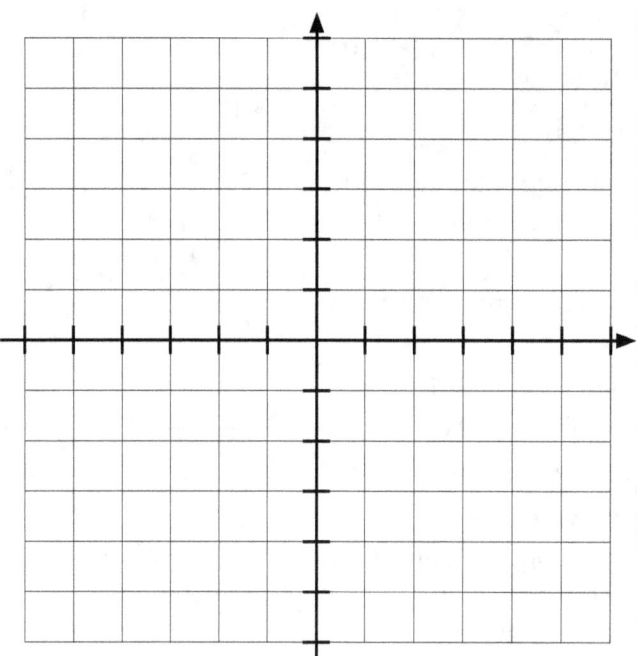

Exercise 6 You Try

Sketch the graph of the parabola, $y = x^2 - 2x - 3$. Be sure to do the following and write all points as ordered pairs.

 a) Determine if the parabola opens upward or downward.

 b) Find the y-intercept.

 c) Find any x-intercepts. If the answer is irrational, round to the nearest hundredth.

 d) Find the vertex.

 e) Find the line of symmetry.

 f) Find the point symmetric to the y-intercept.

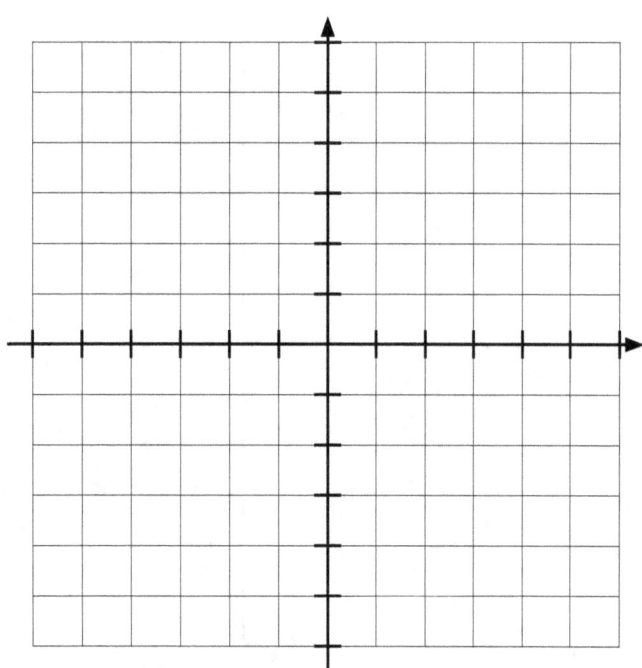

Exercise 7 **Class Example**

Sketch the graph of the parabola, $y = -x^2 + 4$. Be sure to do the following and write all points as ordered pairs.

a) Determine if the parabola opens upward or downward.

b) Find the y-intercept.

c) Find any x-intercepts. If the answer is irrational, round to the nearest hundredth.

d) Find the vertex.

e) Find the line of symmetry.

f) Find the point symmetric to the y-intercept.

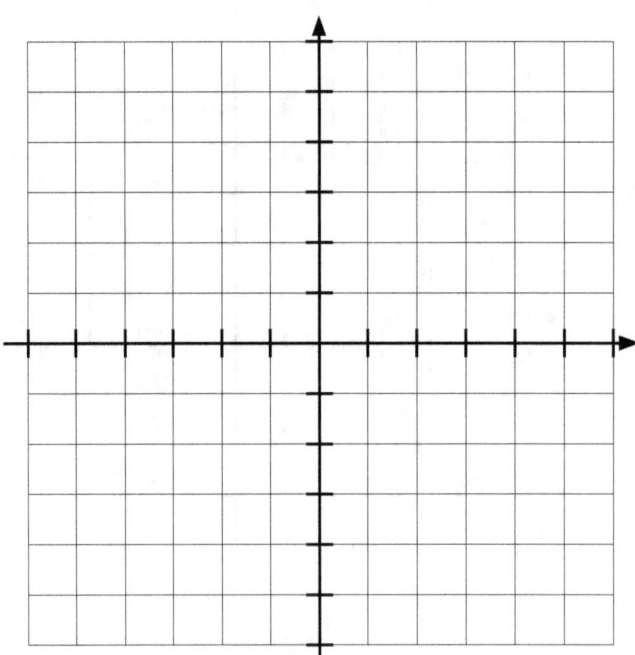

Exercise 8 **You Try**

Sketch the graph of the parabola, $y = -x^2 + 6x$. Be sure to do the following and write all points as ordered pairs.

a) Determine if the parabola opens upward or downward.

b) Find the y-intercept.

c) Find any x-intercepts. If the answer is irrational, round to the nearest hundredth.

d) Find the vertex.

e) Find the line of symmetry.

f) Find the point symmetric to the y-intercept.

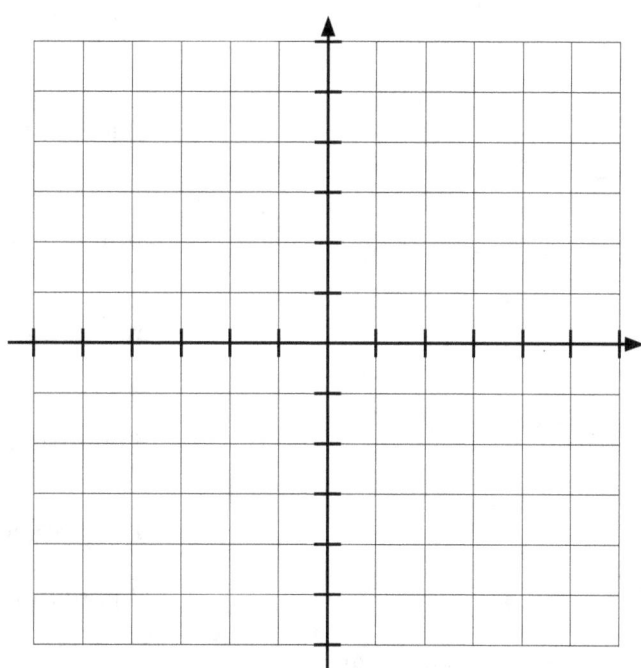

Example 4 Sketch the graph of the parabola, $y = x^2 - 6x + 2$. Be sure to do the following and write all points as ordered pairs.

a) Determine if the parabola opens upward or downward.

b) Find the y-intercept.

c) Find any x-intercepts. If the answer is irrational, round to the nearest hundredth.

d) Find the vertex.

e) Find the line of symmetry.

f) Find the point symmetric to the y-intercept.

Solution.

First identify $a = 1$, $b = -6$ and $c = 2$.

a) The leading coefficient, $a = 1 > 0$. Therefore, the parabola opens upward.

b) To find the y-intercept, let $x = 0$.
$$y = (0)^2 - 6(0) + 2 = 2$$
So, the y-intercept is $(0, 2)$.

c) To find x-intercepts, let $y = 0$ and solve for x. The trinomial is not factorable. In this case, we will solve for x by using the quadratic formula. Then, substitute the values into the quadratic formula.

$$x = \frac{-b \pm \sqrt{b^2 - 4ac}}{2a} \qquad \text{Substitute values}$$

$$x = \frac{-(-6) \pm \sqrt{(-6)^2 - 4(1)(2)}}{2(1)} \qquad \text{Perform indicated operation}$$

$$x = \frac{6 \pm \sqrt{36 - 8}}{2} \qquad \text{Subtract radicand}$$

$$x = \frac{6 \pm \sqrt{28}}{2} \qquad \text{Simplify radical}$$

$$x = \frac{6 \pm 2\sqrt{7}}{2} \qquad \text{Factor GCF in numerator}$$

$$x = \frac{2(3 \pm \sqrt{7})}{2} \qquad \text{Simplify fraction}$$

$$x = 3 \pm \sqrt{7} \qquad \text{Exact solution}$$

$$x \approx 5.65 \text{ or } x \approx 0.35 \qquad \text{Approximate solution}$$

The x-intercepts are $(5.65, 0)$ and $(0.35, 0)$.

d) The x-coordinate of the vertex is $x = -\dfrac{b}{2a} = -\dfrac{-6}{2(1)} = 3$

Substitute $x = 3$ into the quadratic equation to find the y-coordinate of the vertex.
$y = (3)^2 - 6(3) + 2 = 9 - 18 + 2 = -7$
The vertex is $(3, -7)$.

e) The line of symmetry is $x = 3$. It is a vertical line going through the vertex.

f) The y-intercept, $(0, 2)$, is 3 units to the left of the line of symmetry, $x = 3$. By symmetry, there is a point 3 units to the right of $x = 3$. That point is $(6, 2)$ and is symmetric to the y-intercept.

We can now use all these information to graph the parabola.

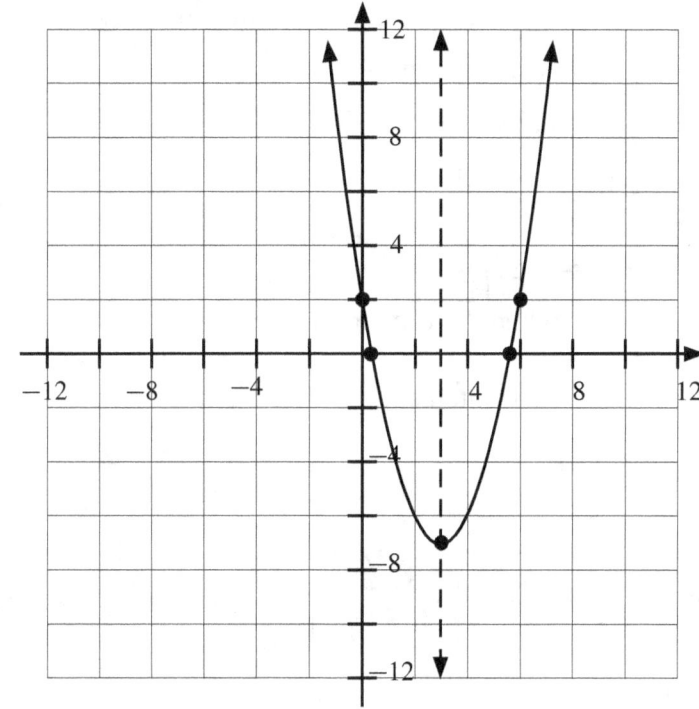

Example 5 Sketch the graph of the parabola, $y = -x^2 + 5$. Be sure to do the following and write all points as ordered pairs.

a) Determine if the parabola opens upward or downward.

b) Find the y-intercept.

c) Find any x-intercepts. If the answer is irrational, round to the nearest hundredth.

d) Find the vertex.

e) Find the line of symmetry.

f) Find the point symmetric to the y-intercept.

Solution.
First identify $a = -1$, $b = 0$ and $c = 5$.

a) The leading coefficient, $a = -1 < 0$. Therefore, the parabola opens downward.

b) To find the y-intercept, let $x = 0$.
   $y = -(0)^2 - 5 = -5$
   So, the y-intercept is $(0, 5)$.

c) To find x-intercepts, let $y = 0$ and solve for x. The trinomial is not factorable. In this case, we will solve for x by using the square root property.

$-x^2 + 5 = 0$	Subtract 5 from each side
$-x^2 = -5$	Multiply each side by -1
$x^2 = 5$	Take the square root of each side
$\sqrt{x^2} = \sqrt{5}$	Simplify
$|x| = \sqrt{5}$	Solve the absolute value equation
$x = \pm\sqrt{5}$	Exact solution
$x \approx 2.24$ or $x \approx -2.24$	Approximate solution

The x-intercepts are $(2.24, 0)$ and $(-2.24, 0)$.

d) The x-coordinate of the vertex is $x = -\dfrac{b}{2a} = -\dfrac{0}{2(1)} = 0$

   Substitute $x = 0$ into the quadratic equation to find the y-coordinate of the vertex.
   $y = -(0)^2 + 5 = 0 + 5 = 5$
   The vertex is $(0, 5)$. This happens to also be the y-intercept.

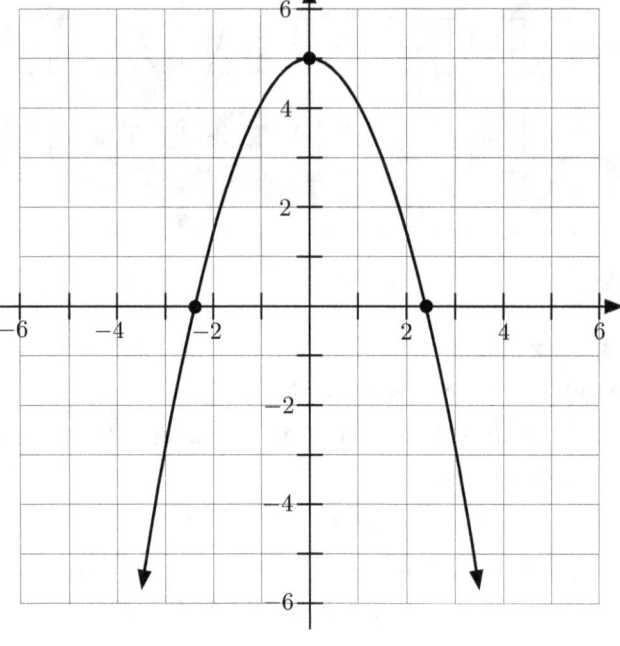

a) The line of symmetry is $x = 0$, the y-axis.

b) The y-intercept, $(0, 3)$, is exactly on the line of symmetry, $x = 0$. It is symmetric to itself.

We can now use all these information to graph the parabola.

Exercise 9 **Class Example**

Sketch the graph of the parabola, $y = x^2 - 8$. Be sure to do the following and write all points as ordered pairs.

a) Determine if the parabola opens upward or downward.

b) Find the y-intercept.

c) Find any x-intercepts. If the answer is irrational, round to the nearest hundredth.

d) Find the vertex.

e) Find the line of symmetry.

f) Find the point symmetric to the y-intercept.

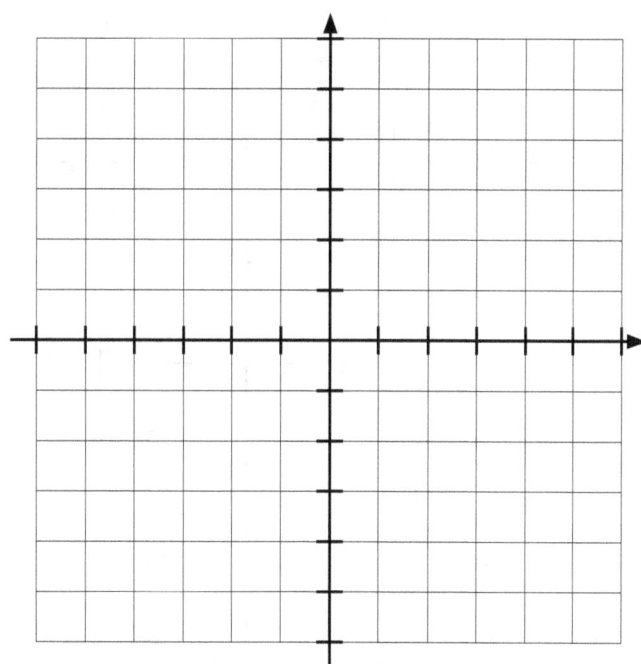

Exercise 10 **Class Example**

Sketch the graph of the parabola, $y = -\dfrac{1}{2}x^2 + 2x + 5$. Be sure to do the following and write all points as ordered pairs.

a) Determine if the parabola opens upward or downward.

b) Find the y-intercept.

c) Find any x-intercepts. If the answer is irrational, round to the nearest hundredth.

d) Find the vertex.

e) Find the line of symmetry.

f) Find the point symmetric to the y-intercept.

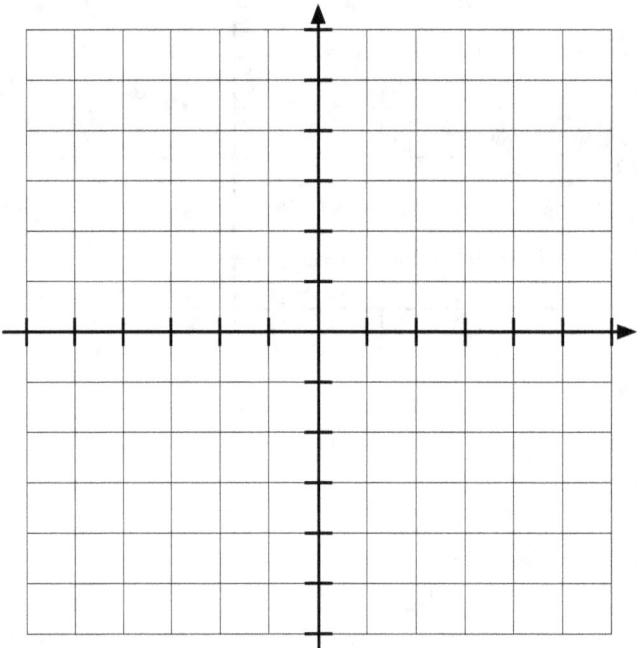

Exercise 11 **You Try**

Sketch the graph of the parabola, $y = x^2 + 4x + 1$. Be sure to do the following and write all points as ordered pairs.

a) Determine if the parabola opens upward or downward.

b) Find the y-intercept.

c) Find any x-intercepts. If the answer is irrational, round to the nearest hundredth.

d) Find the vertex.

e) Find the line of symmetry.

f) Find the point symmetric to the y-intercept.

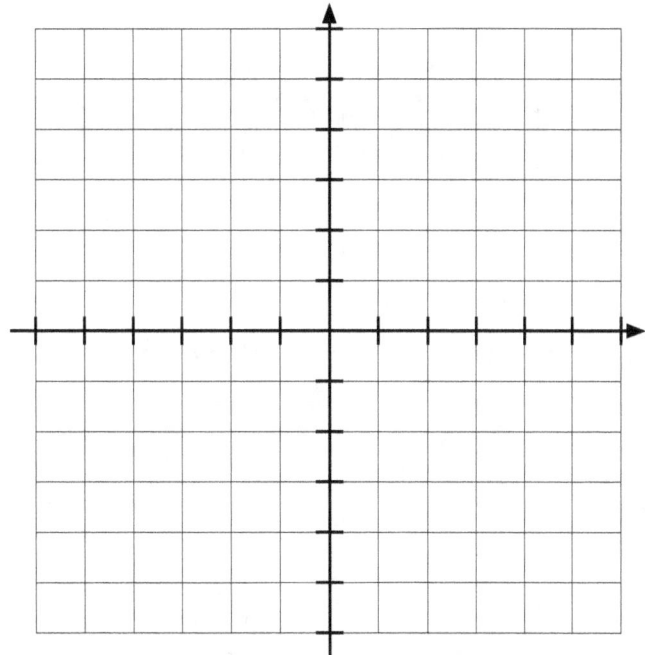

Exercise 12 **You Try**

Sketch the graph of the parabola, $y = -\dfrac{1}{3}x^2 + 6$. Be sure to do the following and write all points as ordered pairs.

a) Determine if the parabola opens upward or downward.

b) Find the y-intercept.

c) Find any x-intercepts. If the answer is irrational, round to the nearest hundredth.

d) Find the vertex.

e) Find the line of symmetry.

f) Find the point symmetric to the y-intercept.

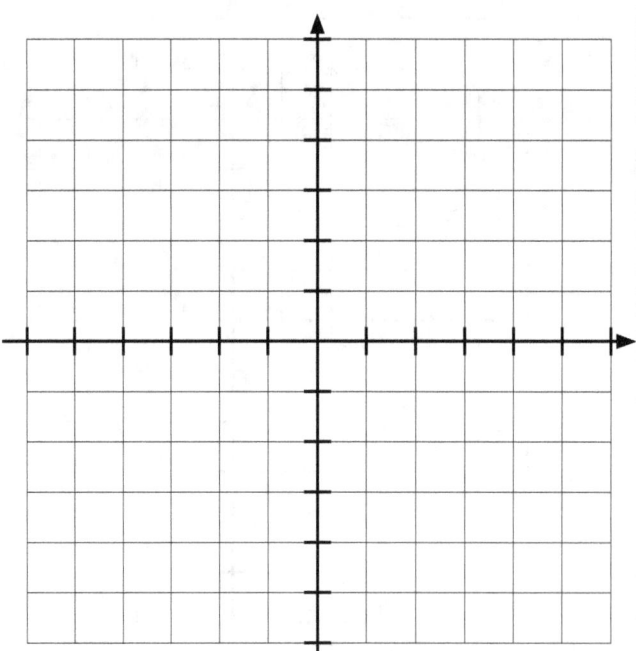

We will now look at how to graph parabolas with one or no x-intercepts. The idea of symmetry will play a big role in graphing these parabolas.

Example 6 Sketch the graph of the parabola, $y = -x^2 + 4x - 4$. Be sure to do the following and write all points as ordered pairs.

a) Determine if the parabola opens upward or downward.

b) Find the y-intercept.

c) Find any x-intercepts. If the answer is irrational, round to the nearest hundredth.

d) Find the vertex.

e) Find the line of symmetry.

f) Find the point symmetric to the y-intercept.

Solution.
First identify $a = -1$, $b = 4$ and $c = -4$.

a) The leading coefficient, $a = -1 < 0$. Therefore, the parabola opens downward.

b) To find the y-intercept, let $x = 0$.
$y = -(0)^2 + 4(0) - 4 = -4$
So, the y-intercept is $(0, -4)$.

c) To find x-intercepts, let $y = 0$ and solve for x. In this case, we can solve for x by factoring.

$$-x^2 + 4x - 4 = 0 \qquad \text{Multiply each term by } -1$$
$$x^2 - 4x + 4 = 0 \qquad \text{Factor trinomial}$$
$$(x - 2)(x - 2) = 0 \qquad \text{Use the zero-product property to solve for } x$$
$$x - 2 = 0 \text{ or } x - 2 = 0$$
$$x = 2 \text{ or } x = 2$$

The solutions are exactly the same. Therefore, we have only 1 x-intercept, $(2, 0)$.

d) The x-coordinate of the vertex is $x = -\dfrac{b}{2a} = -\dfrac{4}{2(-1)} = 2$
Substitute $x = 2$ into the quadratic equation to find the y-coordinate of the vertex.
$y = -(2)^2 + 4(2) - 4 = -4 + 8 - 4 = 0$

The vertex is $(2, 0)$. This happens to also be the x-intercept.

e) The line of symmetry is $x = 2$.

f) The y-intercept, $(0, -4)$, is 2 units to the left of the line of symmetry, $x = 2$. By symmetry, there is a point 2 units to the right of $x = 2$. That point is $(4, -4)$ and is symmetric to the y-intercept.

We can now use these information to graph the parabola.

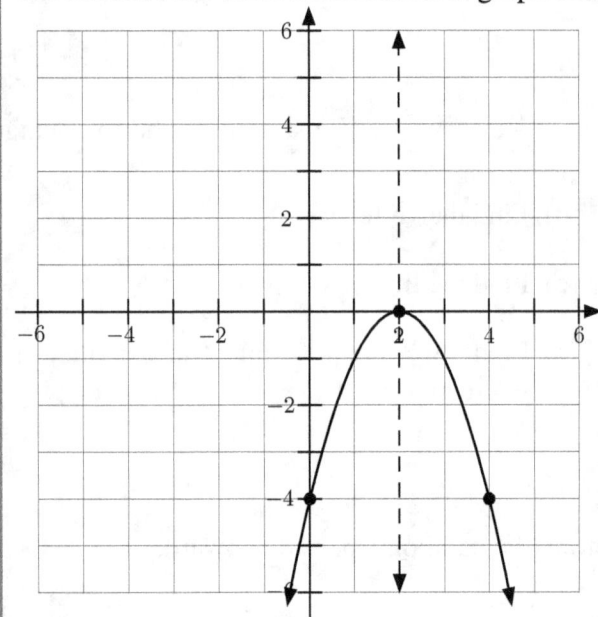

Example 7 Sketch the graph of the parabola, $y = 3x^2 + 4$. Be sure to do the following and write all points as ordered pairs.

a) Determine if the parabola opens upward or downward.

b) Find the y-intercept.

c) Find any x-intercepts. If the answer is irrational, round to the nearest hundredth.

d) Find the vertex.

e) Find the line of symmetry.

f) Find the point symmetric to the y-intercept.

Solution.
First identify $a = 3$, $b = 0$ and $c = 4$.

a) The leading coefficient, $a = 3 > 0$. Therefore, the parabola opens downward.

b) To find the y-intercept, let $x = 0$.
   $y = 3(0)^2 + 4 = 3(0) + 4 = 4$
   So, the y-intercept is $(0, 4)$.

c) To find x-intercepts, let $y = 0$ and solve for x. The trinomial is not factorable. In this case, we will solve for x by using the square root property.

$$3x^2 + 4 = 0$$
$$3x^2 = -4$$
$$x^2 = \frac{-4}{3}$$
$$\sqrt{x^2} = \sqrt{\frac{-4}{3}}$$

$\sqrt{\dfrac{-4}{3}}$ is not a real number. Therefore, there are no x-intercepts.

d) The x-coordinate of the vertex is $x = -\dfrac{b}{2a} = -\dfrac{0}{2(3)} = 0$

Substitute $x = 0$ into the quadratic equation to find the y-coordinate of the vertex.
$y = 3(0)^2 + 4 = 4$
The vertex is $(0,4)$. This also happens to be the y-intercept.

e) The line of symmetry is $x = 0$, the y-axis.

f) The y-intercept, $(0,4)$, is exactly on the line of symmetry, $x = 0$. It is symmetric to itself.

At this point, we have only found 1 point of the parabola. We cannot graph the parabola with only one point. Pick any x-value that is different from the x-coordinate of the vertex. We will choose $x = 1$. Substitute $x = 1$ into the quadratic equation to solve for the y-coordinate.

$y = 3(1)^2 + 4$

$y = 3(1) + 4$

$y = 3 + 4$

$y = 7$

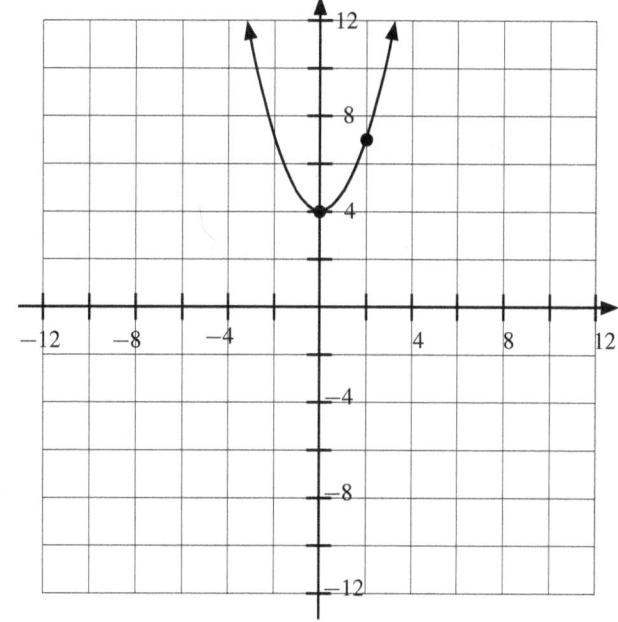

We now have another point on the parabola, $(1,7)$. This point is 1 unit to the right of the line of symmetry, $x = 0$. Using symmetry, we find another point, $(-1,7)$ that is 1 unit to the left of the line of symmetry. We can now use these information to graph the parabola.

Exercise 13 **Class Example**

Sketch the graph of the parabola, $y = -\frac{1}{2}x^2$. Be sure to do the following and write all points as ordered pairs.

a) Determine if the parabola opens upward or downward.

b) Find the y-intercept.

c) Find any x-intercepts. If the answer is irrational, round to the nearest hundredth.

d) Find the vertex.

e) Find the line of symmetry.

f) Find the point symmetric to the y-intercept.

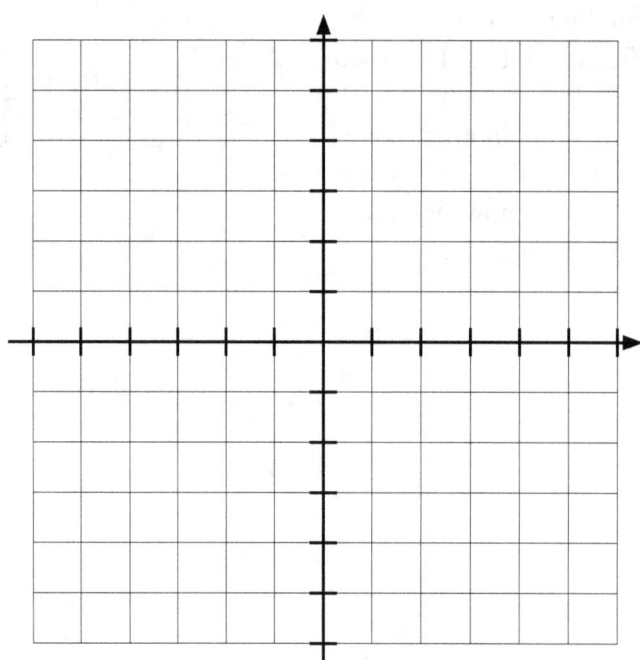

Exercise 14 Class Example

Sketch the graph of the parabola, $y = 2x^2 + 4x + 3$. Be sure to do the following and write all points as ordered pairs.

a) Determine if the parabola opens upward or downward.

b) Find the y-intercept.

c) Find any x-intercepts. If the answer is irrational, round to the nearest hundredth.

d) Find the vertex.

e) Find the line of symmetry.

f) Find the point symmetric to the y-intercept.

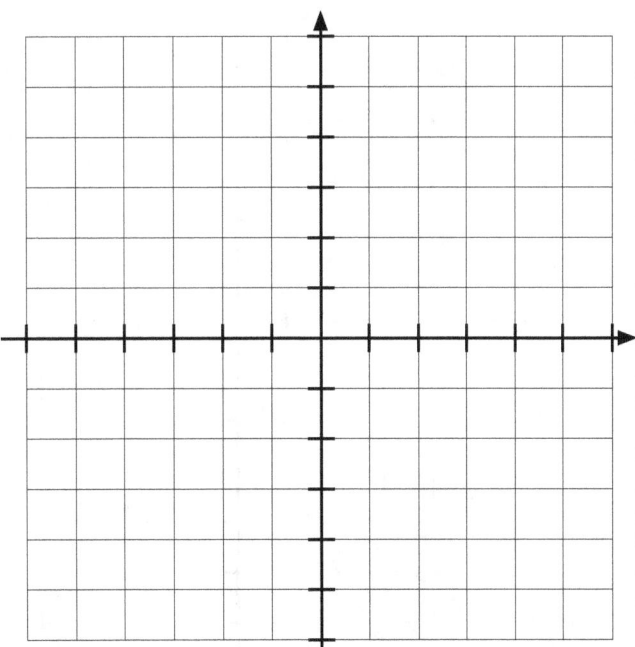

Exercise 15 **You Try**

Sketch the graph of the parabola, $y = 4x^2 + 4x + 1$. Be sure to do the following and write all points as ordered pairs.

a) Determine if the parabola opens upward or downward.

b) Find the y-intercept.

c) Find any x-intercepts. If the answer is irrational, round to the nearest hundredth.

d) Find the vertex.

e) Find the line of symmetry.

f) Find the point symmetric to the y-intercept.

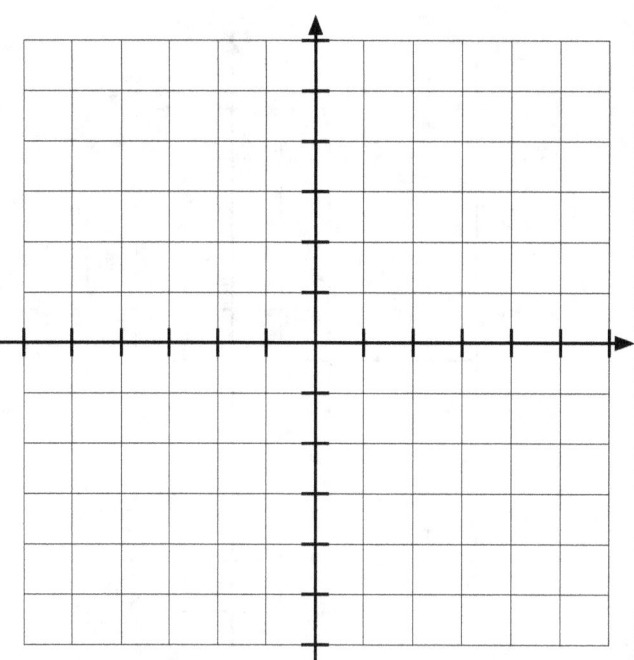

Exercise 16 Class Example

Sketch the graph of the parabola, $y = -2x^2 - 3$. Be sure to do the following and write all points as ordered pairs.

a) Determine if the parabola opens upward or downward.

b) Find the y-intercept.

c) Find any x-intercepts. If the answer is irrational, round to the nearest hundredth.

d) Find the vertex.

e) Find the line of symmetry.

f) Find the point symmetric to the y-intercept.

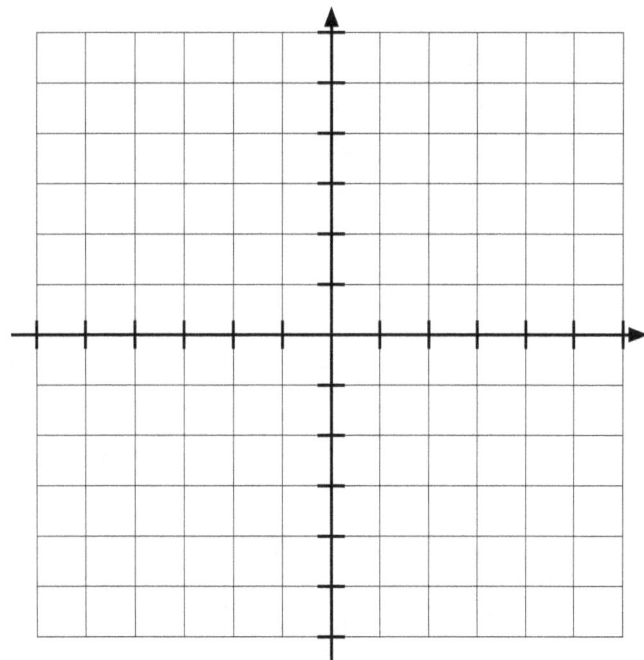

8.5: Exercises

Sketch the graphs of each quadratic equation. Your graph must include:

(a) the y-intercept

(b) any x-intercepts

(c) the vertex

(d) the line of symmetry

1. $y = x^2 + 2x - 3$

2. $y = -x^2 + 4x + 6$

3. $y = -x^2 + 6x - 8$

4. $y = x^2 + 4x + 4$

5. $y = -x^2 - 2$

6. $y = x^2 + 2x + 1$

7. $y = \dfrac{x^2}{2} - 2x$

8. $y = x^2 + 4x$

9. $y = -x^2 + 6x$

10. $y = x^2 - 2x - 8$

8.6   Applications of quadratic Part II

Objective: To solve application problems involving quadratic equations

In this section, we will look at applications involving quadratic equations. When an application problem involves geometry, it is always a good idea to draw a picture first. This will help facilitate in setting up the equation.

Applications Involving Area

Let us first look at some common geometric formulas.

- Perimeter of any polygon = sum of all the sides

- Area of a Rectangle = Length $\times$ Width

- Area of a Triangle = $\dfrac{1}{2}$(Base $\times$ Height)

Example 1 The area of a triangle is 35 square inches. The base of the triangle is 3 inches shorter than the height. Find the dimensions of the triangle. Be sure to include the units in your answers.

Solution.

Let h = height of the triangle

Let $h - 3$ = base of the triangle

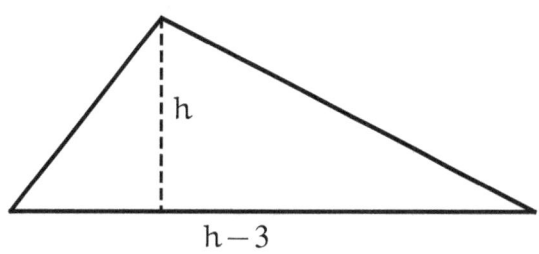

Use this information and the area of a triangle formula to solve for h.

$$\text{Area of the triangle} = \frac{1}{2}(\text{Base} \times \text{Height}) \qquad \text{Area of a triangle formula}$$

$$35 = \frac{1}{2}(h-3)(h) \qquad \text{Substitute information}$$

$$35 = \frac{1}{2}(h-3)(h) \qquad \text{Multiply each side by 2}$$

$$70 = (h-3)(h) \qquad \text{Multiply right-hand side}$$

$$70 = h^2 - 3h \qquad \text{Subtract 70 from each side}$$

$$0 = h^2 - 3h - 70 \qquad \text{Factor trinomial}$$

$$0 = (h-10)(h+7) \qquad \text{Apply zero-product property}$$

$$h - 10 = 0 \text{ or } h + 7 = 0 \qquad \text{Solve for } h$$

$$h = 10 \text{ or } h = -7 \qquad \text{Our Solution}$$

Note that $h = -7$ is an extraneous solution since the height of a triangle can never be a negative value. Therefore, the height of the triangle is 10 inches. Since the base is 3 inches shorter than the height, the base must be 7 inches. Let us confirm that we have the correct solution by calculating the area of the triangle.

$$\text{Area of the triangle} = \frac{1}{2}(\text{Base} \times \text{height}) \qquad \text{Area of a triangle formula}$$

$$35 \overset{?}{=} \frac{1}{2}(7)(10) \qquad \text{Substitute Base} = 7 \text{ and Height} = 10$$

$$35 \overset{?}{=} \frac{1}{2}(70) \qquad \text{Perform indicated operation}$$

$$35 = 35 \quad \checkmark$$

The dimensions of the triangle are 10 inches for the height and 7 inches for the base.

Exercise 1 Class Example
The area of a triangle is 12 square feet. The height of the triangle is 5 feet more than its base. Find the dimensions of the triangle. Be sure to include the units in your answers.

Exercise 2 **Class Example**
If the area of a rectangle is 104 square meters and the perimeter is 42 meters, find the dimensions of the rectangle. Be sure to include the units in your answers.

Exercise 3 **You Try**
The area of a rectangle is 88 square meters. Its width is 3 meters shorter than its length. Find the dimensions of the rectangle.

You Try

If the area of a rectangle is 220.5 square feet and the perimeter is 63 feet, find the dimensions of the rectangle. Be sure to include the units in your answers.

Pythagorean Theorem

The Pythagorean Theorem is used if we are dealing with the sides of a right triangle. In a right triangle, the legs are the two sides that form the right angle. The side opposite the right angle is called the hypotenuse. The hypotenuse is also the longest side of a right triangle.

Pythagorean Theorem

Given a right triangle with legs, a, and b, and hypotenuse, c, the Pythagorean Theorem states that the sum of the squares of the legs of a right triangle is equal to the square of its hypotenuse.

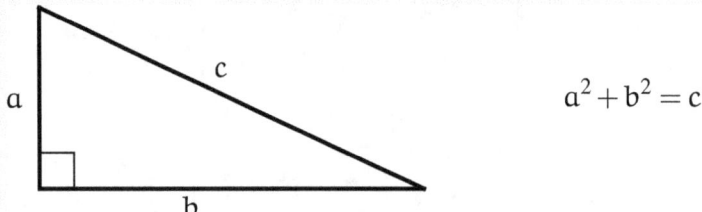

$$a^2 + b^2 = c^2$$

Example 2 A rectangle is 10 inches long and 17 inches wide. What is the length of the diagonal of this rectangle? Give both exact answer and approximate answer, rounded to one decimal place. Be sure to first make a sketch and include the units in your final answer.

Solution.

Let us first draw a picture.

To solve this problem, we will need to use the Pythagorean Theorem, $a^2 + b^2 = c^2$ where $a = 10$ and $b = 17$. Substitute a and b into the Pythagorean Theorem and solve for c.

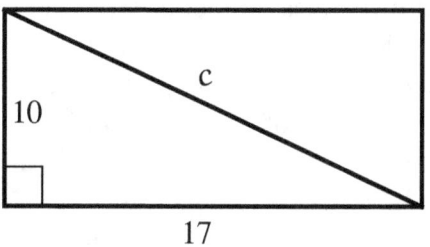

$a^2 + b^2 = c^2$	Substitute $a = 10$ and $b = 17$
$(10)^2 + (17)^2 = c^2$	Perform indicated operation
$100 + 289 = c^2$	Add
$389 = c^2$	Take the square root of each side
$\sqrt{389} = \sqrt{c^2}$	Simplify
$\sqrt{389} = |c|$	Solve the absolute value equation
$c = \pm\sqrt{389}$	Exact answer
$c \approx \pm 19.7$	Approximate answer

$c = -\sqrt{389} \approx -19.7$ is an extraneous solution because the length of the diagonal of a rectangle can never be a negative number.

Therefore, the length of the diagonal of the rectangle is $\sqrt{389}$. This is our exact answer. The diagonal is approximately 19.7 inches.

Exercise 5 **Class Example**
A right triangle has a hypotenuse of 13 in. One leg is 7 inches shorter than the other leg. Find the lengths of the two legs of the right triangle. Be sure to sketch first and include the units in your answers.

Exercise 6 Class Example

A pole stands in a field and the owner wants to secure it with a guy-wire (a wire stretched from some place on the pole down to the ground, pulled away from the base). The owner has 40 feet of guy-wire. He wants to attach the wire to the pole at a certain height that is twice the distance the wire is from the base of the pole. How high above the ground is the guy-wire attached to the pole?

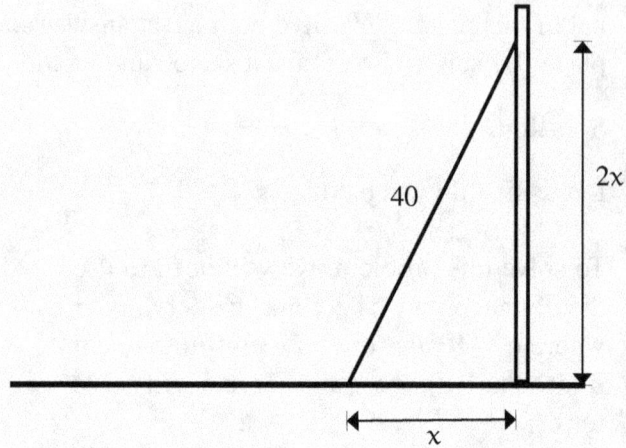

Exercise 7 Class Example

A right triangle has a hypotenuse of 13 in. One leg is 7 inches shorter than the other leg. Find the lengths of the two legs of the right triangle. Be sure to sketch first and include the units in your answers.

Exercise 8 You Try

A right triangle has a hypotenuse of 10 mm. One leg is 2 mm longer than the other leg. Find the lengths of the two legs of the right triangle. Be sure to sketch first include the units in your answers.

Exercise 9 You Try

One leg of a right triangle is 9 inches long. The other leg is 3 inches shorter than the hypotenuse. Find the lengths of the other leg and the hypotenuse of the right triangle. Be sure to sketch first and include the units in your answers.

Maximum-Minimum Problems

In application problems involving quadratic equations, questions regarding maximum or minimum usually involve determining the vertex. Remember that if a quadratic equation is in standard form, $y = ax^2 + bx + c$, the x-coordinate of the vertex is $x = -\frac{b}{2a}$. The y-coordinate of the vertex is obtained by substituting the computed x-value into the quadratic equation.

Example 3 Terry owns a kayaking company that takes people on a tour around San Juan Island. Since Terry is also a whiz at math, he calculates his daily profit, P, to be $P = -5x^2 + 190x - 1200$, where $x \geqslant 0$ is the number of people that go on the kayaking tour daily.

a) One rainy day, Terry had 6 people go on a tour. What was Terry's profit that day?

b) How many people need to go on a tour in order for Terry's business to break even?

c) How many people need to go on a tour so Terry can maximize his daily profit?

d) What is the maximum daily profit?

Solution.

a) To find out Terry's profit when 6 people go on a tour, substitute $x = 6$ into the profit equation and solve for P.

$$P = -5(6)^2 + 190(6) - 1200$$
$$= -180 + 1140 - 1200$$
$$= -240$$

The answer is negative. That means if 6 people go on a kayaking tour, Terry loses $240 that day.

b) Breaking even means there is no gain or loss of money. The business does not make a profit. Set $P = 0$ and solve for x.

$$0 = -5x^2 + 190x - 1200 \qquad \text{Divide each side by } -5$$
$$0 = x^2 - 38x + 240 \qquad \text{Factor trinomial}$$
$$0 = (x - 8)(x - 30) \qquad \text{Use zero product property}$$
$$x - 8 = 0 \text{ or } x - 30 = 0 \qquad \text{Solve for x}$$
$$x = 8 \text{ or } x = 30 \qquad \text{Our Solution}$$

In order for Terry's business to break even, Terry needs either 8 people or 30 people to go on a kayaking tour daily.

c) To find the number of people that needs to go on a tour so Terry can maximize his daily profit, we need to find the vertex. In this equation, we see that the coefficient of x^2 is $a = -5$ and the coefficient of x is $b = 190$. Put this information into the vertex formula to find the x-coordinate.

$$x = -\frac{b}{2a}$$

$$= -\frac{190}{2(-5)}$$

$$= \frac{-190}{-10}$$

$$= 19$$

19 people need to go on the tour each day for Terry to maximize his daily profit.

d) To find the maximum profit, substitute $x = 19$ into the profit equation.

$$P = -5x^2 + 190x - 1200$$

$$= -5(19)^2 + 190(19) - 1200$$

$$= -1805 + 3610 - 1200$$

$$= 605$$

Terry's maximum daily profit is $605. This is achieved when 19 people go on the tour.

Exercise 10 **Class Example**
A local park wants to enclose a rectangular playing area using 100 feet of fencing material. The amount of playing area, A, based on its width, x, is given by the equation: $A = 50x - x^2$.

a) What is the maximum area that can be surrounded with 100 feet of fencing?

b) What is the width of the playing area?

Exercise 11 **You Try**

A manufacturer's daily production costs, C for producing x number of toaster ovens is given by the equation: $C = 0.25x^2 - 11x + 216$.

a) What is the manufacturer's minimum daily production cost?

b) How many toaster ovens can be produced at that cost?

Exercise 12 **Class Example**

A company's daily profit, P (in dollars), for selling x items, is modeled by the following equation: $P = -0.5x^2 + 40x - 300$.

a) What is the company's daily profit if it sells 25 items?

b) How many items must the company sell in order to earn $50 a day?

c) How many items need to be sold to achieve maximum daily profit?

d) What is the maximum daily profit?

Exercise 13 You Try

A model rocket, launched straight up from the ground, reaches a height, h, in feet above the ground according to the equation, $h = -16t^2 + 192t$, where $t > 0$ is the time, in seconds, after the launch.

a) How many seconds will it take until the rocket reaches 300 feet? (Round your answer to the nearest tenth)

b) How high is the rocket after 6 seconds?

c) When will the rocket hit the ground?

d) How many seconds after launch will the rocket reach maximum height?

e) What is the rocket's maximum height?

8.6: Exercises

Solve each of the following application problems. Be sure to show all your work and include units in your answers.

1. A 52-inch TV hasa diagonal of 52 inches. If the height of the TV is 29 inches, how wide is it? Round your answer to the nearest whole number.

2. The hypotenuse of a right triangle is 7 inches long. One leg is 3 inches longer than the other. Find the length of each leg. Round answers to the nearest tenth.

3. The length of a rectangular garden is twice its width. The area of the garden is 72 square yards. Find the dimensions of the garden.

4. When each side of a square is increased by 4 inches, the area becomes 9 times larger than the original square. Find the length of the side of the original square.

5. The length of a rectangular garden is 3 yards longer than the width. The area of the garden is 54 square yards. Find the dimensions of the garden.

6. The infield of a baseball field is a diamond, with a base at each corner. Each corner is a right angle. The distance from home plate to 1st base is 90 feet, as is the distance from 1st to 2nd base, 2nd to 3rd base, and 3rd base to home.

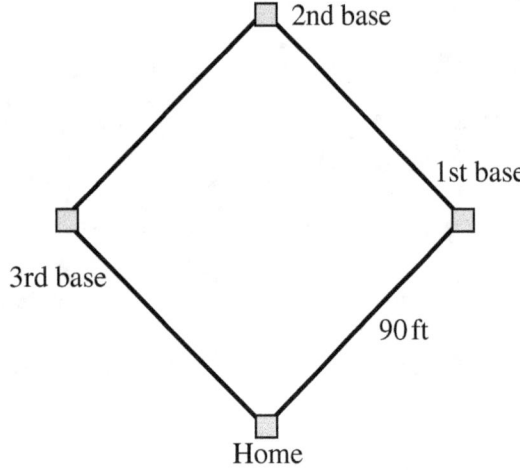

(a) If a ball is thrown from home plate straight to 2nd base, how far will it travel? Give both exact and approximate answers rounded to two decimal places.

(b) If the average time for a ball thrown from home plate straight to 2nd base is 2.0 seconds, at what speed was the ball thrown in feet/second? In miles/hour? Round your answers to the nearest whole number.

7. The cost C of selling x algebra textbooks is $C = \frac{1}{4}x^2 - 35x + 2000$.

   (a) How many algebra textbooks must be sold to minimize cost?

   (b) What is the minimum cost?

8. A local coffee shop has determined that its daily revenue R depends on the price of a cup of coffee p, based on the equation $R = -120p^2 + 684p$.

   (a) If the price of a cup of coffee is $2.00, what is the daily revenue?

(b) If the coffee shop wants to earn \$972 in daily revenue, how much should they charge for a cup of coffee?

(c) What is the coffee shop's maximum daily revenue?

(d) At what price per cup of coffee will maximum daily revenue be achieved?

9. The graph of the height h in feet of a ball t seconds after it has been kicked is shown below. Using the graph, answer the following questions.

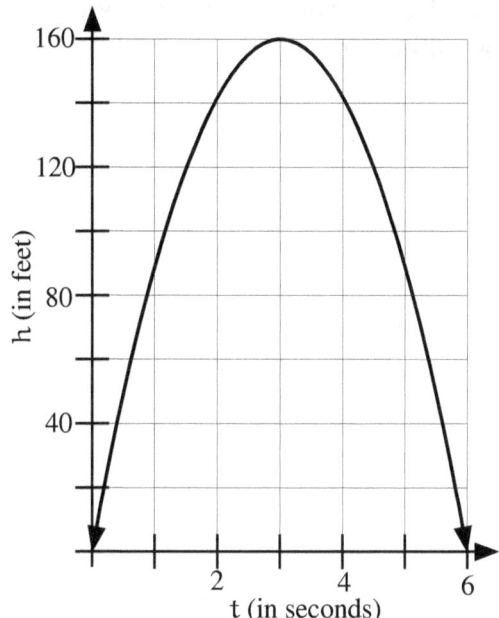

(a) How high did the ball get?

(b) How many seconds did it take to reach that height?

(c) How many seconds did it take for the ball to come back to the ground?

(d) The equation of this graph is $h = -16t^2 + 96t$. Confirm your answers above algebraically.

Chapter 8 Assessment

Simplify the following expressions if possible.

1. $\sqrt{z^2}$

2. $\sqrt{(-4)^2}$

Solve the following equations by the indicated method. Give your answer in exact and simplified form. If the answer is irrational, round to the nearest hundredth.

3. $x^2 = 20$ using the Square Root Property

4. $w^2 + 6w + 4 = 0$ by Completing the Square

5. $2m^2 - 4m + 1 = 0$ using the Quadratic Formula

Solve the following quadratic equations by any method. Give your answer in exact and simplified form. If the answer is irrational, round to the nearest hundredth.

6. $m^2 - 4m = 16$

8. $(3y - 7)^2 = 25$

7. $w^2 - 8w - 3 = 0$

9. $\frac{1}{3}x^2 + \frac{1}{6}x - \frac{1}{2} = 0$

Sketch the graph of each quadratic equation. Be sure to do the following and write all points as an ordered pair.

  a.) **Determine if the parabola opens upward or downward.**

  b.) **Find the y-intercept.**

  c.) **Find any x-intercepts. If answer is irrational, round to the nearest hundredth.**

  d.) **Find the vertex.**

  e.) **Find the line of symmetry.**

10. $y = x^2 - 4x + 3$

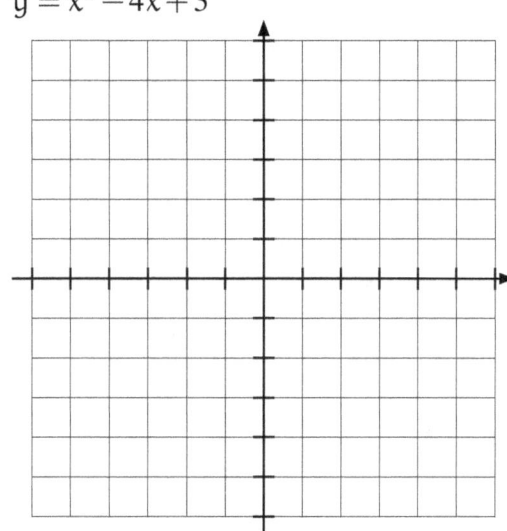

11. $y = -2x^2 + 6$

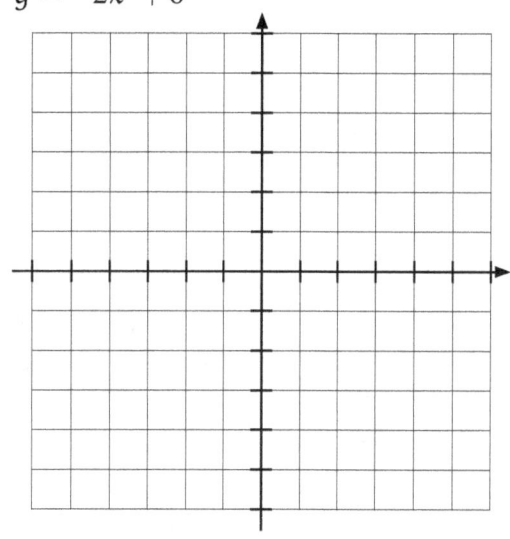

Solve the following word problems.

12. Jack wants to build a flower bed in the shape of an isosceles right triangle (that is, the two legs of the right triangle are the same length). He wants the longest side of the triangle to be 34 feet long. Find the dimensions of the flower bed. Round to the nearest tenth and be sure to include units in your answers.

13. Jenny is designing a banner for Father's day. The banner is 12 feet longer than it is wide and the total area is 64 square feet. Find the length and width of the banner. Be sure to include the units in your answers.

9. Radicals Part II

9.1 Simplifying Radicals

Objective: To be able to simplify expressions involving square roots and cube roots

Previously, we saw how to simplify numerical expressions with square roots. In this section, we will study how to simplify algebraic expressions with different roots. First, let us learn some vocabulary.

Given $\sqrt[n]{a}$, then:

- $\sqrt{}$ **is the radical sign**
- **n is the index**
- **a is the radicand**

The index must always be a positive integer greater than 1. When there is no index written, it is assumed to be 2, which means we have a square root.

Note. Do not confuse $\sqrt[n]{a}$ with $n\sqrt{a}$. They are two different expressions and need to be written properly to avoid confusion. The n in $\sqrt[n]{a}$ is the index of the radical expression. It is a superscript written on the upper left-hand corner of the radical sign. On the other hand, the n in $n\sqrt{a}$ is the coefficient of the square root of a. It means n times the square root of a.

It is also important to say the radical expressions correctly to prevent any misunderstanding. Let us take a look at some examples of radical expressions and how they are properly worded.

Mathematically	*In Words*
$\sqrt{m}$	the square root of m
$\sqrt[3]{n}$	the cube root of n
$4\sqrt{a}$	Four times the square root of a
$5p^2\sqrt[3]{y}$	Five p squared times the cube root of y

How do radicals (or roots) work? Roots perform the inverse operation of applying a positive integer exponent to a base. For example, if we square 7 (or take 7 to the second power), we get 49. If we take the square root of 49, we get 7. The square root is the inverse operation of raising a base to a power of 2. Similarly, if we cube of 4 (or take 4 to the power of 3), we get 64. If we take the cube root of 64, we get 4. The cube root is the inverse operation of raising a base to a power of 3. That is, $\sqrt[n]{a} = b$ because $b^n = a$.

In the example above, you may wonder why $\sqrt{49} \neq -7$ when $(-7)^2 = (-7)(-7) = 49$. In dealing with square roots, we always refer to the principal root. The **principal root** is the positive value of the square root. To denote a negative answer, a negative sign is placed in front of the square root sign. In other words, $\sqrt{49} = 7$ and $-\sqrt{49} = -7$.

Example 1 Simplify the following.

a) $\sqrt{81}$

b) $\sqrt{-4}$

c) $\sqrt[3]{125}$

d) $\sqrt[3]{125}$

Solution.

a) $\sqrt{81} = 9$. Think of a positive number whose square is 81. Since $9^2 = 81$, the square root of 49 must be 7.

b) $\sqrt{-4}$ is a complex number. There is no real number such that when you multiply it by itself will give an answer -4.

c) $\sqrt[3]{125}$. Think of a number whose cube is 125. Since $5^3 = 125$, the cube root of 125 must be 5.

d) $\sqrt[3]{-125} = -5$. Think of a number whose cube is -125. Since $(-5)^3 = -125$, the cube root of -125 must be -5.

Note. For square roots, a negative radicand yields no real number solution. That is because there is no real number such that when you multiply it by itself will yield a negative answer. For cube roots, a negative radicand yields a negative answer. That is because you can multiply a negative number an odd number of times and get a negative result.

Exercise 1 **Class Example**
Simplify the following.

a) $\sqrt{16}$

c) $\sqrt[3]{8}$

b) $\sqrt{-16}$

d) $\sqrt[3]{-8}$

Exercise 2 **You Try**
Simplify the following.

a) $\sqrt{64}$

c) $\sqrt[3]{64}$

b) $\sqrt{-64}$

d) $\sqrt[3]{-64}$

We will not always encounter perfect roots. In simplifying radicals, take note of the index. If the index is 2, you want to find perfect square factors. If the index is 3, you want to find perfect cube factors. In some cases, we will also need to use the product rule for radicals to simplify further.

Product Rule for Radicals

$$\sqrt[n]{ab} = \sqrt[n]{a} \cdot \sqrt[n]{b}$$

(if n is even, then $a \geqslant 0$ and $b \geqslant 0$ for a radical to have a real number solution;
if n is odd, then a and b can be any real number)

Example 2 Simplify the following.

a) $\sqrt{72}$

b) $\sqrt[3]{72}$

Solution.

a) Factor 72 such that one of the factors is a perfect square.

$$\begin{aligned}
\sqrt{72} &= \sqrt{36 \cdot 2} & &\text{Apply product rule for radicals} \\
&= \sqrt{36} \cdot \sqrt{2} & &\text{Simplify perfect root} \\
&= 6\sqrt{2} & &\text{Our Solution}
\end{aligned}$$

b) Factor 72 such that one of the factors is a perfect cube.

$$\sqrt[3]{72} = \sqrt[3]{8 \cdot 9} \qquad \text{Apply product rule for radicals}$$
$$= \sqrt[3]{8} \cdot \sqrt[3]{9} \qquad \text{Simplify perfect root}$$
$$= 2\sqrt[3]{9} \qquad \text{Our Solution}$$

Note. A radical expression is in its simplest form if no perfect root can be factored.

Exercise 3 **Class Example**
Simplify the following.

a) $\sqrt{24}$ b) $\sqrt[3]{24}$

Exercise 4 **Class Example**
Simplify the following.

a) $\sqrt{32}$ b) $\sqrt[3]{32}$

If there is a coefficient in front of the radical sign, the operation between the coefficient and the radical is multiplication.

Example 3 Simplify the following.

a) $-4\sqrt{45}$ b) $7\sqrt[3]{-16}$

Solution.

a) Factor 45 such that one of the factors is a perfect square.

$$-4\sqrt{45} = -4\sqrt{9 \cdot 5} \qquad \text{Apply product rule for radicals}$$
$$= -4\sqrt{9} \cdot \sqrt{5} \qquad \text{Simplify perfect root}$$
$$= -4 \cdot 3 \cdot \sqrt{5} \qquad \text{Multiply coefficients}$$
$$= -12\sqrt{5} \qquad \text{Our Solution}$$

b) Factor -16 such that one of the factors is a perfect cube.

$$7\sqrt[3]{-16} = 7\sqrt[3]{-8 \cdot 2}$$ Apply product rule for radicals

$$= 7\sqrt[3]{-8} \cdot \sqrt[3]{2}$$ Simplify perfect root

$$= 7 \cdot -2 \cdot \sqrt[3]{2}$$ Multiply coefficients

$$= -14\sqrt[3]{2}$$ Our Solution

Exercise 5 Class Example
Simplify the following.

a) $5\sqrt{63}$

b) $4\sqrt[3]{-81}$

Exercise 6 You Try
Simplify the following.

a) $-7\sqrt{12}$

b) $5\sqrt[3]{-54}$

Let us take a look at fractional coefficients.

Example 4 Simplify the following.

a) $\dfrac{3}{7}\sqrt{50}$

b) $\dfrac{\sqrt[3]{-40}}{4}$

Solution.

a) Factor 50 such that one of the factors is a perfect square.

$$\frac{3}{7}\sqrt{50} = \frac{3}{7}\sqrt{25 \cdot 2}$$　　　　　　　Apply product rule for radicals

$$= \frac{3}{7}\sqrt{25} \cdot \sqrt{2}$$　　　　　　　Simplify perfect root

$$= \frac{3}{7} \cdot 5 \cdot \sqrt{2}$$　　　　　　　Multiply coefficients

$$= \frac{15}{7}\sqrt{2}$$　　　　　　　Our Solution

Note. $\frac{15}{7}\sqrt{2}$ can also be written as $\frac{15\sqrt{2}}{7}$

b) Factor -40 such that one of the factors is a perfect cube.

$$\frac{\sqrt[3]{-40}}{4} = \frac{\sqrt[3]{-8 \cdot 5}}{4}$$　　　　　　　Apply product rule for radicals

$$= \frac{\sqrt[3]{-8} \cdot \sqrt[3]{5}}{4}$$　　　　　　　Simplify perfect root

$$= \frac{-2\sqrt[3]{5}}{4}$$　　　　　　　Simplify fraction

$$= \frac{-\sqrt[3]{5}}{2}$$　　　　　　　Our Solution

Note. $\frac{-\sqrt[3]{5}}{2} = -\frac{\sqrt[3]{5}}{2}$ which can also be written as $-\frac{1}{2}\sqrt[3]{5}$

Exercise 7 Class Example
Simplify the following.

a) $\dfrac{\sqrt{27}}{6}$

b) $\dfrac{2}{3}\sqrt[3]{32}$

Exercise 8 **You Try**
Simplify the following.

a) $\dfrac{5}{9}\sqrt{18}$

b) $\dfrac{\sqrt[3]{24}}{6}$

Simplifying Radicals Involving Variables

Variables are sometimes part of the radicand as well. A radical expression is in its simplest form if no variable in the radicand has a higher power than the index. Since variables can take on positive or negative values, we have to be very careful when simplifying. Recall the square root property from the previous chapter which states that $\sqrt{x^2} = |x|$.

What about cube roots? What is $\sqrt[3]{x^3} =$? Let us take a look at the following examples.

$$\text{When } x = 2: \ \sqrt[3]{(2)^3} = \sqrt[3]{8} = 2$$
$$\text{When } x = -2: \ \sqrt[3]{(-2)^3} = \sqrt[3]{-8} = -2$$

From the examples above, we observe that when x is a positive or negative real number, $\sqrt[3]{x^3}$ is equal to x itself. In general, $\sqrt[3]{x^3} = x$.

Example 5 Simplify the following.

a) $\sqrt{y^3}$

b) $\sqrt[3]{m^2}$

Solution.

a) We see that the radicand's power, 3, is higher than the index, 2. Therefore, the radical is not in simplest form. Factor y^3 such that one of the factors is a perfect square.

$$\sqrt{y^3} = \sqrt{y^2 \cdot y} \qquad \text{Apply product rule for radicals}$$
$$= \sqrt{y^2} \cdot \sqrt{y} \qquad \text{Recall that } \sqrt{y^2} = |y|$$
$$= |y|\sqrt{y} \qquad \text{Our Solution}$$

b) The radical expression is already in its simplest form because the radicand's power, 2, is less than the index, 3.

Exercise 9 Class Example
Simplify the following.

a) $\sqrt{a^4}$

c) $\sqrt[3]{x^4}$

b) $-2\sqrt{100c^3}$

d) $\sqrt[3]{-4n^3}$

Exercise 10 You Try
Simplify the following.

a) $\sqrt{36m}$

d) $\sqrt[3]{x^5}$

b) $5\sqrt{y^3}$

e) $\sqrt[3]{-6p^2}$

c) $-3\sqrt{32a^2}$

f) $3\sqrt[3]{-c^3}$

9.1: Exercises

Simplify the following, if possible.

1. $\sqrt{36}$

2. $\sqrt{-36}$

3. $\sqrt[3]{-125}$

4. $\sqrt{28}$

5. $\sqrt[3]{3000}$

6. $\sqrt{98}$

7. $\sqrt[3]{-72}$

8. $7\sqrt{63}$

9. $7\sqrt{-63}$

10. $-7\sqrt{63}$

11. $5\sqrt[3]{16}$

12. $5\sqrt[3]{-16}$

13. $-5\sqrt[3]{-16}$

14. $\dfrac{\sqrt{50}}{10}$

15. $\dfrac{5\sqrt{48}}{4}$

16. $\dfrac{2}{3}\sqrt{54}$

17. $\sqrt{64y^2}$

18. $\sqrt{100n^3}$

19. $-\sqrt{100k^4}$

20. $\sqrt[3]{-64y^3}$

21. $5\sqrt[3]{8h^4}$

22. $-\sqrt[3]{27n^5}$

23. $-7\sqrt{36x^2}$

24. $\sqrt{32p^4}$

25. $-\sqrt{50g^2}$

26. $\sqrt{20a^3}$

27. $\sqrt[3]{20a^3}$

9.2 Adding and Subtracting Radicals

Objective: To add and subtract radicals

Like and Unlike Radical Expressions

The process of adding and subtracting radicals is very similar to that of adding and subtracting polynomials. It is only possible to add or subtract radicals when we have like terms. When radicals are in simplest form, like terms mean that the radicals have the same index and the same radicand. If there is a variable in front of the radical sign, the variables have to be alike and raised to the same power to be considered like terms.

> **Example 1** Determine whether each pair of radical expressions is alike or not. Explain why.
>
> a) $\sqrt{5x}, \sqrt{3x}$
>
> b) $\sqrt{11y}, \sqrt[3]{11y}$
>
> c) $5\sqrt{7}, -8\sqrt{7}$
>
> d) $3x\sqrt{5}, 3y\sqrt{5}$
>
> **Solution.**
>
> a) Unlike expressions because the radicands, 3x and 5x are different.
>
> b) Unlike expressions because the indices are different.
>
> c) Like expressions since both terms have the same radicand, 7, and the same index, 2.
>
> d) Unlike expressions because the variables in front of the radical signs are different.

> Exercise 1 **Class Example**
> Determine whether each pair of radical expressions is alike or not. Explain why.
>
> a) $\sqrt{2a}, \sqrt[3]{2a}$
>
> c) $-\sqrt{3}, 2\sqrt{3}$
>
> b) $\sqrt{5}, \sqrt{10}$
>
> d) $4x\sqrt{7}, 5x^2\sqrt{7}$

Exercise 2 **You Try**
Determine whether each pair of radical expressions is alike or not. Explain why.

a) $5\sqrt{2}, -8\sqrt{3}$

c) $9\sqrt{a}, 9\sqrt{b}$

b) $4\sqrt{y}, 4\sqrt[3]{y}$

d) $6x\sqrt{5}, -2x\sqrt{5}$

Adding and Subtracting Radical Expressions

When adding and subtracting radical expressions, we keep the radical part unchanged while adding and subtracting coefficients, much the same way as combining like terms with polynomials. Let us take a look at the similarity.

Example 2 Combine like terms, if possible, given the following expression.

a) Polynomial expression: $5x + 3x - 2x$
   Radical expression: $5\sqrt{11} + 3\sqrt{11} - 2\sqrt{11}$

b) Polynomial expression: $7x^3 + 5x^2$
   Radical expression: $7\sqrt[3]{x} + 5\sqrt{x}$

Solution.

a) In the polynomial expression, all the terms are alike. Therefore, we can add and subtract the coefficients while leaving the variable unchanged. Similarly, all the terms in the radical expression are alike. We can add and subtract the coefficients while leaving the radical part unchanged.

$$\text{Polynomial expression: } 5x + 3x - 2x = 6x$$
$$\text{Radical expression: } 5\sqrt{11} + 3\sqrt{11} - 2\sqrt{11} = 6\sqrt{11}$$

b) In the polynomial expression, the variable exponents are different. Therefore, the terms are considered unlike and cannot be combined. Similarly, the index of each radical expression is different. The terms are considered to be unlike and cannot be combined.

Exercise 3 **Class Example**
Perform the indicated operation, if possible.

a) $2\sqrt{5}+3\sqrt{5}$

c) $8\sqrt[3]{7}-\sqrt[3]{7}$

b) $7\sqrt{6}-9\sqrt{6}+\sqrt{6}$

d) $5\sqrt{2}-4\sqrt[3]{5}-7\sqrt{2}+\sqrt[3]{2}$

Exercise 4 **You Try**
Perform the indicated operation, if possible.

a) $8\sqrt{3}+2\sqrt{3}$

c) $4\sqrt{5}-7\sqrt[3]{5}$

b) $\sqrt[3]{6}-5\sqrt[3]{6}$

d) $\sqrt{2}-3\sqrt{5}-\sqrt{2}+\sqrt{5}$

Sometimes radical expressions seem to have no like terms. However, if we simplify the expression first, we may find that they do in fact have like terms. Remember to make sure that the radical expression is always in simplified form before adding or subtracting.

Example 3 Perform the indicated operation and write your answer in simplified form.

a) $7\sqrt{20}+\sqrt{80}$

b) $6\sqrt{18}-2\sqrt{98}+\sqrt{50}$

Solution.

a) The radical expressions are not in simplified form. Factor radicands, 10 and 80, such that one of its factors is a perfect square.

$$
\begin{aligned}
7\sqrt{20}+\sqrt{80} &= 7\sqrt{4\cdot 5}+\sqrt{16\cdot 5} && \text{Apply product rule for radicals} \\
&= 7\cdot\sqrt{4}\cdot\sqrt{5}+\sqrt{16}\cdot\sqrt{5} && \text{Simplify perfect roots} \\
&= 7\cdot 2\sqrt{5}+4\sqrt{5} && \text{Multiply coefficients} \\
&= 14\sqrt{5}+4\sqrt{5} && \text{Combine like terms} \\
&= 18\sqrt{5} && \text{Our Solution}
\end{aligned}
$$

b) The radical expressions are not in simplified form. Factor 18, 98 and 50 such that one of its factors is a perfect square

$$6\sqrt{18}-2\sqrt{98}+\sqrt{50}=6\sqrt{9\cdot2}-2\sqrt{49\cdot2}+\sqrt{25\cdot2} \qquad \text{Apply product rule}$$
$$=6\cdot\sqrt{9}\sqrt{2}-2\sqrt{49}\sqrt{2}+\sqrt{25}\sqrt{2} \qquad \text{Simplify perfect roots}$$
$$=6\cdot3\sqrt{2}-2\cdot7\sqrt{2}+5\sqrt{2} \qquad \text{Multiply coefficients}$$
$$=18\sqrt{2}-14\sqrt{2}+5\sqrt{2} \qquad \text{Combine like terms}$$
$$=9\sqrt{2} \qquad \text{Our Solution}$$

Exercise 5 **Class Example**
Perform the indicated operation and write your answer in simplified form.

a) $4\sqrt{32}-\sqrt{200}$

b) $5\sqrt{28}+2\sqrt{7}-\sqrt{63}$

Exercise 6 **You Try**
Perform the indicated operation and write your answer in simplified form.

a) $4\sqrt{54}+3\sqrt{24}$

b) $\sqrt{27}-7\sqrt{3}+2\sqrt{12}$

9.2: Exercises

Simplify.

1. $4\sqrt{3} + \sqrt{3}$

2. $2\sqrt{7} - 5\sqrt{7}$

3. $\sqrt[3]{6} - 4\sqrt[3]{6}$

4. $2\sqrt{7} - \sqrt[3]{7}$

5. $5\sqrt{5} - \sqrt{5} - 3\sqrt{5}$

6. $-2\sqrt{6} - \sqrt{3} - 3\sqrt{6}$

7. $3\sqrt{10} - 5\sqrt[3]{10} - 2\sqrt{10}$

8. $-3\sqrt{2} + 3\sqrt{5} + 7\sqrt{2} - 4\sqrt{5}$

9. $3\sqrt{6} + 3\sqrt{7} + 8\sqrt{6} - 2\sqrt{7}$

10. $\sqrt{11} + 2\sqrt[3]{11} - 2\sqrt{11} - 5\sqrt[3]{11}$

11. $-7\sqrt[3]{6} + 2\sqrt{6} + 7\sqrt[3]{6} - \sqrt{6}$

12. $-\sqrt{2} + 3\sqrt{8}$

13. $\sqrt{12} + 3\sqrt{27}$

14. $\sqrt{54} - 4\sqrt{6}$

15. $2\sqrt{2} - 3\sqrt{18} - \sqrt{2}$

16. $-3\sqrt{27} + 2\sqrt{3} - \sqrt{12}$

17. $-3\sqrt{18} - \sqrt{8} + 2\sqrt{32}$

18. $6\sqrt{5} - \sqrt{20} - \sqrt{45}$

19. $3\sqrt{6} - 3\sqrt{24} + \sqrt{54}$

20. $3\sqrt{8} - \sqrt{12} + \sqrt{48}$

21. $3\sqrt{54} - \sqrt{50} + 2\sqrt{6} + 2\sqrt{8}$

22. $\sqrt{90} - \sqrt{40} - \sqrt{80} + \sqrt{20}$

9.3 Multiplying and Dividing Radicals

Objective: To be able to multiply and divide radicals using the product and quotient rules for radicals

Multiplication of Radicals

Multiplication of radicals can only be performed if the index on all the radicals are the same. If they are, the operation is performed by multiplying coefficients together and multiplying radicands together. This leads us to the more generalized form of the product rule for radicals.

Generalized Form of the Product Rule for Radicals

$$(c \sqrt[n]{a}) \cdot (d \sqrt[n]{b}) = c\,d \sqrt[n]{ab}$$

(if n is even, then $a \geqslant 0$ and $b \geqslant 0$;
if n is odd, then a and b can be any real number)

Example 1 Perform the indicated operation and simplify your answer.

a) $\sqrt{2} \cdot \sqrt{3}$

b) $(-5\sqrt{10}) \cdot (4\sqrt{6})$

c) $\sqrt{20y} \cdot \sqrt{5y}$

d) $(5\sqrt[3]{4}) \cdot (\sqrt[3]{4})$

Solution.

a) Multiply the radicands together. The result is in simpled form.

$$\sqrt{2} \cdot \sqrt{3} = \sqrt{6} \qquad\qquad \text{Our Solution}$$

b) Multiply the coefficients together and then multiply the radicands together. Simplify the result.

$$
\begin{aligned}
(-5\sqrt{10}) \cdot (4\sqrt{6}) &= -5 \cdot 4\sqrt{10 \cdot 6} && \text{Multiply coefficients \& multiply radicands} \\
&= -20\sqrt{60} && \text{Factor 60 with one factor a perfect square} \\
&= -20\sqrt{4 \cdot 15} && \text{Apply product rule for radicals} \\
&= -20 \cdot \sqrt{4} \cdot \sqrt{15} && \text{Simplify perfect root} \\
&= -20 \cdot 2\sqrt{15} && \text{Multiply coefficients} \\
&= -40\sqrt{15} && \text{Our Solution}
\end{aligned}
$$

c) Multiply coefficients together and then simplify the result.

$$\sqrt{20y} \cdot \sqrt{5y} = \sqrt{100y^2} \qquad \text{Multiply coefficients together}$$
$$= \sqrt{100}\sqrt{y^2} \qquad \text{Apply product rule for radicals}$$
$$= 10|y| \qquad \text{Our Solution}$$

Recall that $\sqrt{y^2} = |y|$.

d) Multiply the coefficients together and then multiply the radicands together. Simplify the result.

$$(5\sqrt[3]{4}) \cdot (\sqrt[3]{4}) = 5\sqrt[3]{4 \cdot 4} \qquad \text{Multiply coefficients and multiply radicands}$$
$$= 5\sqrt[3]{16} \qquad \text{Factor 16 with one factor a perfect cube}$$
$$= 5\sqrt[3]{8 \cdot 2} \qquad \text{Apply product rule for radicals}$$
$$= 5 \cdot \sqrt[3]{8} \cdot \sqrt[3]{2} \qquad \text{Simplify perfect root}$$
$$= 5 \cdot 2\sqrt[3]{2} \qquad \text{Multiply coefficients}$$
$$= 10\sqrt[3]{2} \qquad \text{Our Solution}$$

Exercise 1 **Class Example**
Perform the indicated operation and simplify your answer.

a) $\sqrt{7} \cdot \sqrt{14}$

c) $\sqrt{3n} \cdot \sqrt{12n^2}$

b) $(7\sqrt{6}) \cdot (2\sqrt{12})$

d) $\sqrt[3]{9} \cdot \sqrt[3]{6}$

Exercise 2 You Try

Perform the indicated operation and simplify your answer.

a) $\sqrt{12} \cdot \sqrt{2}$

c) $\sqrt{8a} \cdot \sqrt{2a}$

b) $(-3\sqrt{5}) \cdot (-\sqrt{10})$

d) $\sqrt[3]{3} \cdot \sqrt[3]{18}$

Let us take a look at how to apply the distributive property involving radicals.

Example 2 Perform the indicated operation and simplify your answer.

a) $\sqrt{6}(3\sqrt{10} - \sqrt{15})$

b) $(6 + \sqrt{7})(3 - 2\sqrt{7})$

Solution.

a) Apply the distributive property and simplify the result.

$$
\begin{aligned}
\sqrt{6}(3\sqrt{10} - \sqrt{15}) &= 3\sqrt{6 \cdot 10} - \sqrt{6 \cdot 15} && \text{Multiply radicands} \\
&= 3\sqrt{60} - \sqrt{90} && \text{Find perfect square factors} \\
&= 3\sqrt{4 \cdot 15} - \sqrt{9 \cdot 10} && \text{Apply product rule for radicals} \\
&= 3\sqrt{4}\sqrt{15} - \sqrt{9}\sqrt{10} && \text{Simplify perfect root} \\
&= 3 \cdot 2\sqrt{15} - 3\sqrt{10} && \text{Multiply coefficients} \\
&= 6\sqrt{15} - 3\sqrt{10} && \text{Our Solution}
\end{aligned}
$$

b) Remember to apply the distributive property when multiplying two binomials.

$$(6+\sqrt{7})(3-2\sqrt{7})$$ Multiply binomials

$$=6\cdot3-6\cdot2\sqrt{7}+3\sqrt{7}-2\sqrt{7\cdot7}$$ Multiply coefficients and radicands

$$=18-12\sqrt{7}+3\sqrt{7}-2\sqrt{49}$$ Simplify perfect root

$$=18-12\sqrt{7}+3\sqrt{7}-2\cdot7$$ Multiply last term

$$=18-12\sqrt{7}+3\sqrt{7}-14$$ Combine like terms

$$=4-9\sqrt{7}$$ Our Solution

Exercise 3 Class Example

Perform the indicated operation and simplify your answer.

a) $\sqrt{5}(4-\sqrt{10})$ c) $(2+\sqrt{6})(2-\sqrt{6})$

b) $2\sqrt{8}\,(5\sqrt{2}+\sqrt{3})$ d) $(5-2\sqrt{3})(4+\sqrt{6})$

Exercise 4 **You Try**

Perform the indicated operation and simplify your answer.

a) $2\sqrt{6}\,(3+\sqrt{2})$

c) $(\sqrt{7}-\sqrt{3})(\sqrt{7}+\sqrt{3})$

b) $\sqrt{10}\,(7\sqrt{2}-\sqrt{3})$

d) $(5+\sqrt{3})^2$

Now that we know arithmetic operations on radicals, let us take a look at how to verify our answer if we have an irrational answer.

Example 3 Check if $x = -2\sqrt{3}$ is a solution to the equation $x^2 - 12 = 0$.

Solution.

$$(-2\sqrt{3})^2 - 12 \overset{?}{=} 0 \qquad\qquad \text{Substitute } x = -2\sqrt{3}$$

$$(-2\sqrt{3})(-2\sqrt{3}) - 12 \overset{?}{=} 0 \qquad\qquad \text{Expand square term and multiply}$$

$$4\sqrt{9} - 12 \overset{?}{=} 0 \qquad\qquad \text{Simplify perfect root}$$

$$4\cdot 3 - 12 \overset{?}{=} 0 \qquad\qquad \text{Multiply}$$

$$12 - 12 = 0 \ \checkmark$$

Therefore, $x = -2\sqrt{3}$ is a solution to $x^2 - 12 = 0$.

Exercise 5 Class Example
Check if $x = 1 - 2\sqrt{5}$ is a solution to the equation $x^2 - 2x = 19$.

Exercise 6 You Try

a) Check if $x = 5\sqrt{2}$ is a solution to the equation $x^2 - 50 = 0$.

b) Check if $x = 2 + \sqrt{3}$ is a solution to the equation $x^2 + 1 = 4x$.

Division of Radicals

A radical is considered not simplified if there is a fraction underneath the radical sign or if there is a radical in the denominator. To simplify, we will use a more generalized form of the quotient rule for radicals.

Generalized Form of the Quotient Rule for Radicals

$$\frac{a\sqrt[n]{b}}{c\sqrt[n]{d}} = \frac{a}{c}\sqrt[n]{\frac{b}{d}}$$

(if n is even, then $a \geqslant 0$ and $b \geqslant 0$;
if n is odd, then a and b can be any real number)

Example 4 Perform the indicated operation and simplify your answer.

a) $\dfrac{\sqrt{48}}{\sqrt{27}}$ b) $11\sqrt[3]{\dfrac{54}{2}}$

Solution.

a) Since both radicands, 48 and 27, have a common factor, 3, simplify first.

$$\dfrac{\sqrt{48}}{\sqrt{27}} = \sqrt{\dfrac{48}{27}}$$ Apply quotient rule for radicals and simplify fraction

$$= \sqrt{\dfrac{16}{9}}$$ Apply quotient rule for radicals

$$= \dfrac{\sqrt{16}}{\sqrt{9}}$$ Simplify perfect root

$$= \dfrac{4}{3}$$ Our Solution

b) Since 54 and 2 have a common factor, 2, simplify the fraction first.

$$11\sqrt[3]{\dfrac{54}{2}} = 11\sqrt[3]{27}$$ Simplify fraction and perfect root

$$= 11 \cdot 3$$ Multiply

$$= 33$$ Our Solution

Exercise 7 Class Example

Perform the indicated operation and simplify your answer.

a) $\dfrac{\sqrt{45}}{\sqrt{20}}$

c) $\dfrac{10\sqrt[3]{81}}{\sqrt[3]{24}}$

b) $\dfrac{9}{6}\sqrt{\dfrac{50}{8}}$

d) $\dfrac{1}{5}\sqrt[3]{\dfrac{-192}{3}}$

Exercise 8 You Try

Perform the indicated operation and simplify your answer.

a) $12\sqrt{\dfrac{25}{9}}$

c) $\dfrac{\sqrt[3]{54}}{\sqrt[3]{-16}}$

b) $\dfrac{4\sqrt{75}}{15\sqrt{3}}$

d) $\dfrac{5}{8}\sqrt[3]{\dfrac{40}{5}}$

Rationalizing the Denominator

In most math texts, it is customary to clear the denominator of any radicals. The process is called rationalizing the denominator. Here, we will only focus on clearing the square root term in the denominator. If the radical in the denominator is a monomial and the radicand is not a perfect square, we rationalize the denominator by multiplying both numerator and denominator by the same square root term.

Example 5 Simplify each of the following.

a) $\dfrac{2}{\sqrt{5}}$

b) $\dfrac{\sqrt{15}}{\sqrt{24}}$

Solution.

a) To simplify the expression, we will rationalize the denominator by multiplying both numerator and denominator by $\sqrt{5}$.

$$\frac{2}{\sqrt{5}} = \frac{2}{\sqrt{5}} \cdot \frac{\sqrt{5}}{\sqrt{5}} \qquad \text{Multiply}$$

$$= \frac{2\sqrt{5}}{\sqrt{25}} \qquad \text{Simplify perfect root}$$

$$= \frac{2\sqrt{5}}{5} \qquad \text{Our Solution}$$

b) Both radicals have the same index and both radicands have a common factor, 3. Simplify first before rationalizing the denominator.

$$\frac{\sqrt{15}}{\sqrt{24}} = \sqrt{\frac{15}{24}}$$ Apply quotient rule for radicals and simplify fraction

$$= \sqrt{\frac{5}{8}}$$ Apply quotient rule for radicals

$$= \frac{\sqrt{5}}{\sqrt{8}}$$ Factor 8 with one factor a perfect square

$$= \frac{\sqrt{5}}{\sqrt{4 \cdot 2}}$$ Apply product rule for radicals

$$= \frac{\sqrt{5}}{\sqrt{4} \cdot \sqrt{2}}$$ Simplify perfect root

$$= \frac{\sqrt{5}}{2\sqrt{2}}$$ Rationalize the denominator

$$= \frac{\sqrt{5}}{2\sqrt{2}} \cdot \frac{\sqrt{2}}{\sqrt{2}}$$ Multiply

$$= \frac{\sqrt{10}}{2 \cdot \sqrt{4}}$$ Simplify perfect root

$$= \frac{\sqrt{10}}{2 \cdot 2}$$ Multiply denominator

$$= \frac{\sqrt{10}}{4}$$ Our Solution

Exercise 9 **Class Example**

Perform the indicated operation and simplify your answer.

a) $\dfrac{3}{\sqrt{2}}$

c) $\dfrac{\sqrt{10}}{\sqrt{14}}$

b) $\dfrac{9}{2\sqrt{3}}$

d) $\sqrt{\dfrac{11}{50}}$

Exercise 10 **You Try**

Perform the indicated operation and simplify your answer.

a) $\sqrt{\dfrac{16}{5}}$

c) $\dfrac{\sqrt{12}}{\sqrt{18}}$

b) $\sqrt{\dfrac{1}{2}}$

d) $\dfrac{4\sqrt{7}}{\sqrt{8}}$

9.3: Exercises

Perform the indicated operation and simplify

1. $\sqrt{5} \cdot \sqrt{3}$

2. $-2\sqrt{10} \cdot \sqrt{5}$

3. $(9\sqrt[3]{-4}) \cdot (2\sqrt[3]{6})$

4. $(3\sqrt{5}) \cdot (4\sqrt{6})$

5. $(-4\sqrt[3]{9}) \cdot (\sqrt[3]{6})$

6. $\sqrt{12m} \cdot \sqrt{3m}$

7. $\sqrt{5r} \cdot \sqrt{20r^2}$

8. $\sqrt{2y} \cdot \sqrt{8y}$

9. $\sqrt{6}\left(\sqrt{2}+2\right)$

10. $\sqrt{10}\left(\sqrt{5}+\sqrt{2}\right)$

11. $5\sqrt{3}\left(3+\sqrt{6}\right)$

12. $\left(2+2\sqrt{2}\right)\left(3+\sqrt{2}\right)$

13. $(2-\sqrt{3})(5+2\sqrt{3})$

14. $(4-\sqrt{7})(4+\sqrt{7})$

15. $(2\sqrt{6}+1)^2$

16. $\left(\sqrt{11}+\sqrt{6}\right)\left(\sqrt{11}-\sqrt{6}\right)$

17. $(3-\sqrt{7})^2$

18. $(2\sqrt{3}+\sqrt{5})(5\sqrt{3}+4)$

19. $\dfrac{\sqrt{12}}{\sqrt{3}}$

20. $\dfrac{4\sqrt{15}}{\sqrt{3}}$

21. $\dfrac{4\sqrt{125}}{\sqrt{5}}$

22. $\dfrac{\sqrt{12}}{5\sqrt{100}}$

23. $\sqrt[3]{\dfrac{15}{64}}$

24. $\dfrac{3\sqrt[3]{10}}{5\sqrt[3]{27}}$

25. $\dfrac{2}{\sqrt{3}}$

26. $\dfrac{\sqrt{10}}{\sqrt{6}}$

27. $\sqrt{\dfrac{1}{6}}$

28. $\dfrac{\sqrt{2}}{3\sqrt{5}}$

Verify that x is a solution to the given equation.

29. $x = -5\sqrt{2}$; Equation: $x^2 - 50 = 0$

30. $x = 2 - \sqrt{2}$; Equation: $x^2 - 4x + 2 = 0$

31. $x = -3 + \sqrt{11}$; Equation: $x^2 + 6x = 2$

9.4 Square Root Equations

Objective: To be able to solve square root equations and check for extraneous solutions

Consider an equation with a variable under the square root such as $\sqrt{x} = 4$. We call this a square root equation. To solve for x, we need to isolate the radical expression.

Recall from previous chapter, $(\sqrt{b})^2 = b$, for $b \geqslant 0$. Therefore, to solve the equation, $\sqrt{x} = 4$ for x, we must square each side of the equation.

Example 1 Solve $\sqrt{x} = 3$ for x.

Solution.

$$\sqrt{x} = 3$$ Square each side
$$(\sqrt{x})^2 = (3)^2$$ Perform the indicated operation
$$x = 9$$ Our Solution

Verify that we have the correct solution by substituting $x = 9$ into the original equation.

$$\sqrt{x} = 3$$
$$\sqrt{9} = 3 \ \checkmark$$

Example 2 Solve $\sqrt{y-4} = 5$ for y.

Solution.

$$\sqrt{y-4} = 5$$ Square each side
$$(\sqrt{y-4})^2 = (5)^2$$ Perform the indicated operation
$$y - 4 = 25$$ Add 4 to each side
$$y = 29$$ Our Solution

Verify that we have the correct solution by substituting $y = 29$ into the original equation.

$$\sqrt{y-4} = 5$$
$$\sqrt{29-4} \overset{?}{=} 5$$
$$\sqrt{25} = 5 \ \checkmark$$

Example 3 Solve $\sqrt{2n} - 1 = 7$ for n.

Solution.

Before we square each side, we must first isolate the radical expression.

$$\sqrt{2n} - 1 = 7$$ Add 1 to each side of the equation

$$\sqrt{2n} = 8$$ Square each side

$$(\sqrt{2n})^2 = (8)^2$$ Perform the indicated operation

$$2n = 64$$ Divide each side by 2

$$n = 32$$ Our Solution

Verify that we have the correct solution by substituting $n = 32$ into the original equation.

$$\sqrt{2n} - 1 = 7$$

$$\sqrt{2(32)} - 1 \overset{?}{=} 7$$

$$\sqrt{64} - 1 \overset{?}{=} 7$$

$$8 - 1 = 7 \;\checkmark$$

Example 4 Solve $\sqrt{x+2} = -6$ for x.

Solution.

Looking closely at the problem, you may notice that it has no solution. Why? Because the positive square root of a number can never be negative. However, if you do not recognize that right away, you can always discover this *when you check your answer.*

$$\sqrt{x+2} = -6$$ Square each side

$$(\sqrt{x+2})^2 = (-6)^2$$ Perform the indicated operation

$$x + 2 = 36$$ Subtract 2 from each side

$$x = 34$$ Our Solution

Verify that we have the correct solution by substituting $x = 34$ into the original equation.

$$\sqrt{x+2} = -6$$

$$\sqrt{(34)+2} \overset{?}{=} -6$$

$$\sqrt{36} \neq -6$$

$x = 34$ does not solve the equation. Therefore, the equation has no solution. We say $x = 34$ is an *extraneous solution*.

> **Remark.** An *extraneous solution* is an answer that emerges from the process of solving the problem, but is not a valid solution. When solving square root equations, it is very important to always verify that the answer is a valid solution to the given problem.

Steps in solving a square root problem.

1. Isolate the radical expression.

2. Square each side of the equation.

3. Solve for the variable.

4. Verify that the answer is a valid solution to the given problem.

Exercise 1 **Class Example**
Solve the following equations.

a) $\sqrt{m} = 8$

c) $-\sqrt{5k + 10} = -4$

b) $\sqrt{x} - 4 = 2$

d) $5 + \sqrt{p - 1} = 3$

Exercise 2 **You Try**
Solve the following equations.

a) $\sqrt{w} = 9$

c) $\sqrt{m + 8} + 10 = 2$

b) $-3\sqrt{5x} = -30$

d) $\dfrac{\sqrt{6y + 4}}{3} = 2$

What if there are 2 radical expressions in an equation? We still isolate one of the radical expressions before squaring each side of the equation. Let us take a look at an example.

Example 5 Solve $\sqrt{3x-8} - \sqrt{x} = 0$ for x.

Solution.

$$\sqrt{3x-8} - \sqrt{x} = 0 \qquad \text{Isolate the radicals}$$
$$\sqrt{3x-8} = \sqrt{x} \qquad \text{Square each side}$$
$$(\sqrt{3x-8})^2 = (\sqrt{x})^2 \qquad \text{Perform the indicated operations}$$
$$3x-8 = x \qquad \text{Subtract 3x on each side to solve for x}$$
$$-8 = -2x \qquad \text{Divide each side by } -2$$
$$4 = x \qquad \text{Our Solution}$$

Verify that we have the correct solution by substituting $x = 4$ into the original equation.

$$\sqrt{3x-8} - \sqrt{x} \overset{?}{=} 0$$
$$\sqrt{3(4)-8} - \sqrt{4} \overset{?}{=} 0$$
$$\sqrt{12-8} - \sqrt{4} \overset{?}{=} 0$$
$$\sqrt{4} - \sqrt{4} = 0 \checkmark$$

Exercise 3 Class Example
Solve $\sqrt{3x+8} = \sqrt{4x+2}$ for x.

Exercise 4 You Try

Solve $\sqrt{x} = \sqrt{5x-4}$ for x.

In solving square root equations, we sometimes encounter the square of a binomial. To expand the square of a binomial, remember to always rewrite the expression as a product of two binomials and then multiply them by applying the distributive property. Let us review how to expand the square of a binomial such as $(x-3)^2$.

$$
\begin{aligned}
(x-3)^2 &= (x-3)(x-3) \\
&= x(x-3) - 3(x-3) \\
&= x^2 - 3x - 3x + 9 \\
&= x^2 - 6x + 9
\end{aligned}
$$

Example 6 Solve $x = 5 + \sqrt{4x+1}$ for x.

Solution.

$x = 5 + \sqrt{4x+1}$	Isolate the radical
$x - 5 = \sqrt{4x+1}$	Square each side
$(x-5)^2 = (\sqrt{4x+1})^2$	Perform the indicated operation
$x^2 - 10x + 25 = 4x + 1$	Solve quadratic equation
$x^2 - 14x + 24 = 0$	Factor trinomial
$(x-12)(x-2) = 0$	Apply zero product rule
$x - 12 = 0$ or $x - 2 = 0$	Solve for x
$x = 12$ or $x = 2$	Our Solution

Verify that we have the correct solution by substituting each value into the original equation *one at a time*.

Check $x = 12$:   Check $x = 2$:

$$x = 5 + \sqrt{4x + 1}$$

$$12 \overset{?}{=} 5 + \sqrt{4(12) + 1}$$

$$12 \overset{?}{=} 5 + \sqrt{48 + 1}$$

$$12 \overset{?}{=} 5 + \sqrt{49}$$

$$12 = 5 + 7 \quad \checkmark$$

$$x = 5 + \sqrt{4x + 1}$$

$$12 \overset{?}{=} 5 + \sqrt{4(2) + 1}$$

$$12 \overset{?}{=} 5 + \sqrt{8 + 1}$$

$$12 \overset{?}{=} 5 + \sqrt{9}$$

$$12 \neq 5 + 3$$

$x = 2$ is an extraneous solution. Therefore, $x = 12$ is our solution to the given equation.

Exercise 5 Class Example
Solve the following equations.

a) $\sqrt{6n + 7} = n + 2$

b) $y + \sqrt{10 - y} = 4$

Exercise 6 **You Try**
Solve the following equations.

a) $x = \sqrt{x+6}$

c) $\sqrt{m+11} = m-1$

b) $\sqrt{y}+2 = y$

d) $2+\sqrt{1-8p} = p$

Application Problems

We will now look at different applications involving square root equations. Let us start with the distance between 2 points. The distance formula is derived from the Pythagorean Theorem.

Distance Between Two Points

Given 2 points, (x_1,y_1) and (x_2,y_2), the distance between these two points is given by the formula:

$$d = \sqrt{(x_2-x_1)^2 + (y_2-y_1)^2}$$

Example 7 Find the distance between $(1,0)$ and $(-2,4)$.

Solution.

Let $(x_1, y_1) = (1,0)$ and $(x_2, y_2) = (-2,4)$.

$$d = \sqrt{(x_2 - x_1)^2 + (y_2 - y_1)^2}$$ Distance Formula

$$d = \sqrt{(-2-1)^2 + (4-0)^2}$$ Substitute values and subtract

$$d = \sqrt{(-3)^2 + (4)^2}$$ Square each term

$$d = \sqrt{9 + 16}$$ Add

$$d = \sqrt{25}$$ Take square root

$$d = 5$$ Our Solution

Therefore, $(1,0)$ and $(-2,4)$ are 5 units apart.

Exercise 7 Class Example

Use the distance formula to answer the following questions.

a) Find the distance between point $(5,-3)$ and point $(7,1)$. Give the exact answer in simplified form. If the answer is irrational, provide an approximation, rounded to the nearest tenth.

b) Find the distance between point $(4\sqrt{2}, -\sqrt{3}$ and point $3\sqrt{2}, \sqrt{3})$. Give the Give the exact answer in simplified form. If the answer is irrational, provide an approximation, rounded to the nearest hundredth.

Exercise 8 **You Try**

Use the distance formula to answer the following questions.

a) Find the distance between point $(-4, 1)$ and point $(-6, 3)$. Give the exact answer in simplified form. If the answer is irrational, provide an approximation, rounded to the nearest tenth.

b) Find the distance between point $(\sqrt{5}, \sqrt{2}$ and point $3\sqrt{5}, -\sqrt{2})$. Give the exact answer in simplified form. If the answer is irrational, provide an approximation, rounded to the nearest hundredth.

c) Challenge: Find 2 points on the x-axis that are 5 units aways from the point (6,3).

Exercise 9 **Class Example**

A simple pendulum consists of a string, cord or wire that allows a suspended mass to swing back and forth. The categorization of "simple" comes from the fact that all of the mass of the pendulum is concentrated in its "bob" or suspended mass. If the bob is pulled to the right and then released, the time for it to swing to the left and then back to its original position, known as the period, can be found with the following formula, $T = 2\pi\sqrt{\frac{1}{g}}$, where T is the period in seconds, l is the length of the string in meters, and $g = 9.81$ m/sec^2 is the earth's gravity.

a) If the bob of a pendulum is suspended from a one-meter string, what is its period? Round your answer to the nearest integer.

b) If you want to double the period, how long must the string be? Round your answer to the nearest integer.

Exercise 10 You Try

The formula $s = k\sqrt{d}$ relates the speed, s (in mph) of a car and the distance, d (in feet) of the skid when a driver hits the brakes. k is a constant that changes for different road conditions.

a) What is the speed of a car, if the skid is 124 feet long, on an icy pavement where $k = 25$?

b) If a car is traveling at 55 mph on wet pavement where $k = 3.24$, how far will the car skid after the brakes are applied? Round to the nearest foot.

9.4: Exercises

Solve.

1. $\sqrt{x-4} = 11$

2. $\sqrt{10-y} + 2 = 8$

3. $10 + \sqrt{5x+1} = 3$

4. $\sqrt{7p+2} = 4$

5. $\sqrt{6w+1} - 7 = 0$

6. $-2\sqrt{3c+1} + 15 = 9$

7. $\sqrt{6x-5} - x = 0$

8. $2x = \sqrt{9x-2}$

9. $\sqrt{y+2} = y$

10. $n = 1 + \sqrt{7-n}$

11. $3 + m = \sqrt{6m+13}$

12. $\sqrt{3-3g} = 2g + 1$

13. $h + \sqrt{4h+1} = 5$

14. $\sqrt{x+2} - x = 2$

15. $\sqrt{3k+22} = \sqrt{14-k}$

16. $\sqrt{1-3p} - \sqrt{3p+7} = 0$

17. $\sqrt{3m} = \sqrt{\dfrac{m}{3}+8}$

Word Problems.

18. Find the distance between the following points. Be sure to simplify your answers. If an answer is irrational, approximate to the nearest hundredth.

    (a) $(0,7)$ and $(-3,1)$

    (b) $\left(\sqrt{5}, -\sqrt{2}\right)$ and $\left(-\sqrt{5}, \sqrt{2}\right)$

19. A car is travelling 70 mph on dry pavement. Use the formula $s = k\sqrt{d}$, where s is the car speed (in mph), d is the distance (in feet) of the skid when a driver hits the brakes and $k = 5.34$, to find how far the car skids. Round to the nearest foot.

Challenge Problems.

20. Find 2 points on the y-axis that are 13 units away from the point $(-5, 10)$.

21. Solve the following equations.

    (a) $\sqrt{y+9} = 3 + \sqrt{y}$

    (b) $\sqrt{c+1} = 1 + \sqrt{c-6}$

Rescue Roody!

22. Help Roody figure out his misunderstanding. Roody was asked to solve the following equation: $\sqrt{x+4}-2=x$. Why was Roody's answer marked incorrect?

$$\sqrt{x+4}-2=x$$
$$\sqrt{x+4}=(x+2)^2$$
$$x+4=x^2+4$$
$$0=x^2-x$$
$$0=x(x-1)$$
$$x=0 \text{ or } x=1$$

Check.

When $x=0$:

$$\sqrt{(0)+4}-2\overset{?}{=}0$$
$$\sqrt{4}-2=0 \checkmark$$

When $x=1$:

$$\sqrt{(1)+4}-2\overset{?}{=}1$$
$$\sqrt{5}-2\neq 1$$

Therefore, $x=0$ is the solution to the equation.

Chapter 9 Assessment

Simplify the following.

1. $-\sqrt{20}$

3. $\sqrt{100y^2}$

2. $\sqrt[3]{-80}$

4. $\sqrt[3]{64y^3}$

Perform the indicated operation and simplify your answer.

5. $\sqrt[3]{10}+4\sqrt[3]{10}$

9. $(2\sqrt{10})(\sqrt{6})$

6. $7\sqrt{5}+2\sqrt{6}-9\sqrt{5}-\sqrt{6}$

10. $\sqrt{2}(3+\sqrt{10})$

7. $6\sqrt{18}-\sqrt{50}$

11. $(5+\sqrt{7})(5-\sqrt{7})$

8. $\sqrt{12}+5\sqrt{48}-\sqrt{27}$

12. $(4-\sqrt{6})^2$

Rationalize the denominator.

13. $\dfrac{3}{\sqrt{6}}$

14. $\sqrt{\dfrac{5}{8}}$

Verification.

15. Verify that $x=-3\sqrt{5}$ is a solution to the equation $x^2-75=0$

16. Verify that $x=4+\sqrt{3}$ is a solution to the equation $x^2-8x=13$.

Solve the following equations.

15. $\sqrt{x+5}=7$

17. $\sqrt{n}=n$

16. $\sqrt{2a+5}+8=3$

18. $\sqrt{3k-4}=\sqrt{k+3}$

Word Problem.

19. Use the distance formula, $d=\sqrt{(x_2-x_1)^2+(y_2-y_1)^2}$ to find the distance between point $(7,-1)$ and point $(5,-3)$. Give the exact answer in simplified form. If the answer is irrational, provide an approximation, rounded to the nearest hundredth.

10. Rational Expressions

10.1 Introduction to Rational Expressions

Objective: To evaluate and simplify rational expression

Recall that a **rational number (or fraction)** is a real number that can be expressed as a ratio, $\dfrac{p}{q}$, where p and q are integers and $q \neq 0$. Here are some examples of rational numbers: $7, 0, \frac{2}{3}$ and $-\frac{9}{5}$.

Generalizing this notion, a **rational expression** is an algebraic expression that can be expressed as a ratio, $\dfrac{P}{Q}$, where P and Q are *polynomials* and $Q \neq 0$. Here are some examples of rational expressions: $z^3, 7, 0, -\dfrac{3}{y^2}$ and $\dfrac{2x+4}{x-1}$.

Note. Every rational number is in fact a rational expression.

Rational Expressions: Ratios of Polynomials
Ex. $8z^2, \; -\dfrac{11}{7y}, \; \dfrac{4x^2}{x+9}, \; ...$
Rational Numbers: Ratios of Integers
Ex. $8, \; -\dfrac{11}{7}, \; \dfrac{4}{9}, \; ...$

Evaluating Rational Expressions

Just as we evaluate other types of algebraic expressions, we can evaluate rational expressions. Remember to use parenthesis appropriately.

Example 1 Evaluate the rational expression $\dfrac{3x^2 - 8}{x - 2}$ for the given values.

a) $x = 3$ \qquad\qquad b) $x = -4$ \qquad\qquad c) $x = 2$

Solution.
For each problem, substitute the given value of x and simplify the resulting expression using the order of operation.

a) When $x = 3$:

$$\frac{3x^2 - 8}{x - 2} = \frac{3(3)^2 - 8}{(3) - 2}$$

$$= \frac{3(9) - 8}{1}$$

$$= 27 - 8$$

$$= 19$$

b) When $x = -4$:

$$\frac{3x^2 - 8}{x - 2} = \frac{3(-4)^2 - 8}{(-4) - 2}$$

$$= \frac{3(16) - 8}{-4 - 2}$$

$$= \frac{48 - 8}{-6}$$

$$= \frac{40}{-6}$$

$$= \frac{20}{-3}$$

Note. Since $\dfrac{-a}{b} = \dfrac{a}{-b} = -\dfrac{a}{b}$, we can write $\dfrac{20}{-3}$ as $-\dfrac{20}{3}$ or $\dfrac{-20}{3}$.

c) When $x = 2$:

$$\frac{3x^2 - 8}{x - 2} = \frac{3(2)^2 - 8}{(2) - 2}$$

$$= \frac{3(4) - 8}{2 - 2}$$

$$= \frac{12 - 8}{2 - 2}$$

$$= \frac{4}{0}$$

Since it is not permissible to divide by 0, we conclude that the expression $\dfrac{3x^2 - 8}{x - 2}$ is **undefined** when $x = 2$.

Exercise 1 Class Example

Evaluate the rational expression, $\dfrac{x + 1}{x^2 - 6x + 8}$ for the given values.

a) $x = 1$ c) $x = -1$

b) $x = 0$ d) $x = 4$

Exercise 2 **You Try**

Evaluate the rational expression, $\dfrac{3-y}{y^2-9}$ for the given values.

a) $x = 0$ c) $x = -3$

b) $x = -2$ d) $x = 3$

Determining When a Rational Expression is Undefined

As we saw in the previous examples, a rational expression is undefined for values of the variable that make the denominator 0. Thus, to find all values for which a rational expression is undefined, we simply set the denominator equal to 0 and solve the resulting equation.

Example 2 For what values of the variable is the rational expression $\dfrac{x+3}{x^2-2x}$ undefined?

Solution.

To find the values that make the rational expression undefined, we set the denominator to 0 and solve the resulting equation for x.

$$x^2 - 2x = 0 \qquad\qquad \text{Factor the binomial}$$
$$x(x-2) = 0 \qquad\qquad \text{Use zero-product property}$$
$$x = 0 \text{ or } x - 2 = 0 \qquad\qquad \text{Solve for x}$$
$$x = 0 \text{ or } x = 2 \qquad\qquad \text{Our Solution}$$

We conclude that the rational expression, $\dfrac{x+3}{x^2-2x}$ is undefined when $x = 0$ or $x = 2$.

Exercise 3 Class Example

For what values of the variable is the given expression undefined?

a) $\dfrac{3}{x-7}$

c) $\dfrac{4y}{y^2+5y+6}$

b) $\dfrac{1-p}{2p}$

d) $\dfrac{n-2}{n^2+1}$

Exercise 4 You Try

For what values of the variable is the given expression undefined?

a) $\dfrac{2-x}{x-2}$

c) $\dfrac{z-4}{4}$

b) $\dfrac{5+y}{y^2}$

d) $\dfrac{3t-1}{2t^2-5t-3}$

Simplifying Rational Expressions

Let us recall how to simplify rational numbers (or fractions) by considering the fraction $\dfrac{8}{12}$. First, we find the prime factorization of each number.

$$\frac{8}{12} = \frac{2\cdot 2\cdot 2}{3\cdot 2\cdot 2} \qquad \text{Find prime factorization of 8 and 12}$$

$$= \frac{2\cdot \overset{1}{\cancel{2}}\cdot \overset{1}{\cancel{2}}}{3\cdot \underset{1}{\cancel{2}}\cdot \underset{1}{\cancel{2}}} \qquad \text{Divide common factors}$$

$$= \frac{2}{3} \qquad \text{Simplified fraction}$$

As noted in the beginning of this section, a rational expression is really just a more general type of fraction. Thus, to simplify a rational expression, we use the exact same reasoning as in the above example. That is, we first factor the numerator and denominator completely. Then we seek common factors between the numerator and denominator and simplify by dividing these common factors.

Steps in Simplifying Rational Expressions:

1. *Factor* both numerator and denominator completely

2. Simplify the rational expression by dividing common factors

Example 3 Simplify each of the following rational expressions.

a) $\dfrac{4x}{x^2}$ d) $\dfrac{p^2-2p}{p-2}$

b) $\dfrac{3m^2}{6m-24}$ e) $\dfrac{y+3}{(y+3)^2}$

c) $\dfrac{2n+6}{n^2-9}$ f) $\dfrac{z+7}{z}$

Solution.

a)

$$\frac{4x}{x^2} = \frac{4\cdot x}{x\cdot x} \qquad \text{Factor numerator and denominator completely}$$

$$= \frac{4\cdot \cancel{x}^{1}}{x\cdot \cancel{x}^{1}} \qquad \text{Divide common factors}$$

$$= \frac{4}{x} \qquad \text{Our Solution}$$

b)

$$\frac{3m^2}{6m-24} = \frac{3\cdot m\cdot m}{6\cdot (m-4)} \qquad \text{Factor numerator \& denominator completely}$$

$$= \frac{\cancel{3}^{1}\cdot m\cdot m}{\cancel{6}_{2}\cdot (m-4)} \qquad \text{Divide common factors}$$

$$= \frac{m^2}{2(m-4)} \qquad \text{Our Solution}$$

Note. We could have distributed the factor 2 in the denominator giving us $\dfrac{m^2}{2(m-2)} =$

$\dfrac{m^2}{2m-4}$. While both answers are acceptable, it is recommended that answers be left in their factored form.

c)

$$\frac{2n+6}{n^2-9} = \frac{2\cdot(n+3)}{(n+3)\cdot(n-3)} \qquad \text{Factor numerator \& denominator completely}$$

$$= \frac{2\cancel{(n+3)}^{\,1}}{\cancel{(n+3)}^{\,1}\cdot(n-3)} \qquad \text{Divide common factors}$$

$$= \frac{2}{n-3} \qquad \text{Our Solution}$$

d)

$$\frac{p^2-2p}{p-2} = \frac{p\cdot(p-2)}{p-2} \qquad \text{Factor numerator \& denominator completely}$$

$$= \frac{p\cdot\cancel{(p-2)}^{\,1}}{\cancel{(p-2)}^{\,1}} \qquad \text{Divide common factors}$$

$$= \frac{p}{1} \qquad \text{Simplify fraction}$$

$$= p \qquad \text{Our Solution}$$

e)

$$\frac{y+3}{(y+3)^2} = \frac{y+3}{(y+3)\cdot(y+3)} \qquad \text{Factor denominator}$$

$$= \frac{\cancel{(y+3)}^{\,1}}{\cancel{(y+3)}^{\,1}\cdot(y+3)} \qquad \text{Divide common factors}$$

$$= \frac{1}{y+3} \qquad \text{Our Solution}$$

Note. The answer is not $y+3$. The answer is a fraction with a 1 in the numerator.

f) For the expression, $\dfrac{z+7}{z}$, the numerator and denominator are fully factored and there are no common factors amongst the numerator and denominator. So, the expression is already in simplified form.

Warning. When simplifying rational expressions, it is crucial that you **first factor** both numerator and denominator completely and then simplify the expression by dividing *common factors*. For example, consider the expression, $\dfrac{6}{3x+7}$. If we attempt to simplify this expression by dividing the *term* 6 in the numerator and the *term* 3x in the denominator, we run into problem such as this:

$$\frac{6}{3x+7} = \frac{2}{x+7}$$

However, these are not equivalent expressions. Consider evaluating the two expressions using the value $x = 4$.

$$\frac{6}{3x+7} \overset{?}{=} \frac{2}{x+7}$$

$$\frac{6}{3(4)+7} \overset{?}{=} \frac{12}{(4)+7}$$

$$\frac{6}{12+7} \overset{?}{=} \frac{2}{4+7}$$

$$\frac{6}{19} \neq \frac{2}{11}$$

We see that $\dfrac{6}{3x+7} \neq \dfrac{2}{x+7}$ for $x = 4$. Try other values of x. In fact, there is no value of x that will make the two expressions equal.

For the expression, $\dfrac{6}{3x+7}$, it is a mistake to divide the numerator and denominator by 3 because **3 is not a factor of the denominator!** The factors of the numerator are 2 and 3, whereas the denominator has only the single factor $3x+7$. Thus, there are no common factors between the numerator and the denominator. Therefore, the expression, $\dfrac{6}{3x+7}$ is already in simplified form.

Exercise 5 Class Example

Simplify each of the following rational expressions.

a) $\dfrac{3a}{6a^3}$

d) $\dfrac{x^2-4}{x^2+16}$

b) $\dfrac{m-4}{(m-4)^2}$

e) $\dfrac{y^2-3y}{2y}$

c) $\dfrac{2}{w-6}$

f) $\dfrac{n-3}{n^2-n-6}$

Exercise 6 You Try

Simplify each of the following rational expressions.

a) $\dfrac{2w^4}{6w^2}$

d) $\dfrac{v}{v+8}$

b) $\dfrac{9(z+7)}{6(z+7)^2}$

e) $\dfrac{-n-4}{n+4}$

c) $\dfrac{t+3}{t^2+6t+9}$

f) $\dfrac{y^2-5y}{y^2-25}$

One property of real numbers is that addition is commutative. For example, $3+2=2+3$, $-9+6=6+-9$ and $0+7=7+0$. Similarly, $a+b=b+a$. For rational expressions, if the numerator is the reverse order of the denominator and the operation between terms is addition, then the rational expression simplifies to 1 because of the commutative property of addition. For example,

$$\frac{3+2}{2+3}=\frac{5}{5}=1$$

$$\frac{-9+6}{6+-9}=\frac{-3}{-3}=1$$

$$\frac{0+7}{7+0}=\frac{7}{7}=1$$

$$\frac{a+b}{b+a}=1$$

$$\frac{3x+1}{1+3x}=1$$

This is not true for subtraction because subtraction is not commutative. Let us take a look at some examples.

$$\frac{3-2}{2-3}=\frac{1}{-1}=-1$$

$$\frac{-9-6}{6-(-9)}=\frac{-15}{15}=-1$$

$$\frac{0-7}{7-0}=\frac{-7}{7}=-1$$

We see that the numerator and denominator are negatives of one another. The key to simplifying rational expressions where the numerator is the reverse order of the denominator and the operation between terms is subtraction is to use the fact that

$$b-a=-(a-b)$$

Example 4 Simplify each of the following rational expressions.

a) $\dfrac{9-y}{y-9}$
b) $\dfrac{8+x}{x^2+8x}$
c) $\dfrac{6-3w}{w^2-4}$

Solution.

a)

$$\frac{9-y}{y-9} = \frac{-(y-9)}{y-9} \qquad \text{Use the fact that } 9-y = -(y-9)$$

$$= \frac{-\cancel{(y-9)}^{1}}{\cancel{y-9}^{1}} \qquad \text{Divide common factors}$$

$$= \frac{-1}{1} \qquad \text{Simplify}$$

$$= -1 \qquad \text{Our Solution}$$

Note. We could have factored -1 from the denominator instead of the numerator. We will still get the same answer. Let us take a look.

$$\frac{9-y}{y-9} = \frac{(9-y)}{-(9-y)} \qquad \text{Use the fact that } y-9 = -(9-y)$$

$$= \frac{\cancel{(9-y)}^{1}}{-\cancel{(9-y)}^{1}} \qquad \text{Divide common factors}$$

$$= \frac{1}{-1} \qquad \text{Simplify}$$

$$= -1 \qquad \text{Our Solution}$$

b)

$$\frac{8+x}{x^2+8x} = \frac{8+x}{x(x+8)} \qquad \text{Factor denominator}$$

$$= \frac{x+8}{x(x+8)} \qquad \text{Addition is commutative}$$

$$= \frac{\cancel{(x+8)}^{1}}{x\cancel{(x+8)}^{1}} \qquad \text{Divide common factors}$$

$$= \frac{1}{x} \qquad \text{Our Solution}$$

c)

$$\frac{6-3w}{w^2-4} = \frac{3(2-w)}{(w+2)(w-2)}$$ Factor numerator and denominator

$$= \frac{-3(w-2)}{(w+2)(w-2)}$$ Use the fact that $2-w=-(w-2)$

$$= \frac{-3\cancel{(w-2)}^{\,1}}{(w+2)\cancel{(w-2)}^{\,1}}$$ Divide common factors

$$= \frac{-3}{w+2}$$ Our Solution

Exercise 7 Class Example

Simplify each of the following rational expressions.

a) $\dfrac{3-g}{g-3}$

c) $\dfrac{x^2-9}{18-6x}$

b) $\dfrac{10+2y}{3y+15}$

d) $-\dfrac{1-a}{a-1}$

Exercise 8 You Try

Simplify each of the following rational expressions.

a) $\dfrac{m-5}{5-m}$

c) $\dfrac{y-7}{y^2-7y}$

b) $\dfrac{3x+6}{2+x}$

d) $-\dfrac{z-9}{9-z}$

10.1: Exercises

Evaluate the expression for the given value. Simplify your answers.

1. $\dfrac{v+2}{v}$ when $v = 4$

2. $\dfrac{3-b}{b-9}$ when $b = -3$

3. $\dfrac{x^2-3}{x^2-5x}$ when $x = -1$

4. $\dfrac{4-a^2}{a+4}$ when $a = -2$

5. $\dfrac{b+2}{b^2+b-12}$ when $b = 3$

6. $\dfrac{n^2-n-6}{n-3}$ when $n = 0$

Find all values of the variable that make the given rational expression undefined.

7. $\dfrac{k}{k+10}$

8. $\dfrac{n^2}{10n+5}$

9. $\dfrac{m+7}{3}$

10. $\dfrac{m^2-16}{2m}$

11. $\dfrac{r^2+2}{r^2+1}$

12. $\dfrac{x-1}{1-x}$

13. $\dfrac{2}{p^2-6p}$

14. $\dfrac{2n^2+6n}{n^2-9}$

15. $\dfrac{4a}{a^2-5a-6}$

16. $\dfrac{y-2}{3y^2-y-10}$

Simplify the given rational expression.

17. $\dfrac{4n}{12n^2}$

18. $\dfrac{32x^3}{8x^5}$

19. $\dfrac{h+5}{5+h}$

20. $\dfrac{a-7}{a+7}$

21. $\dfrac{20}{4p+2}$

22. $\dfrac{r-10}{r}$

23. $-\dfrac{m-3}{3-m}$

24. $\dfrac{8m+16}{2+m}$

25. $\dfrac{n-9}{81-9n}$

26. $\dfrac{x+1}{x^2+8x+7}$

27. $\dfrac{4x^2+28x}{4x^2}$

28. $\dfrac{-5b-20}{5b+20}$

29. $\dfrac{n^2+2n+1}{6+6n}$

30. $\dfrac{3a-10}{10+3a}$

31. $\dfrac{54+9v}{v^2-4v-60}$

33. $\dfrac{k^2-12k+32}{64-k^2}$

32. $\dfrac{b^2+14b+48}{b^2+15b+56}$

34. $\dfrac{x^2-5x+4}{3x^2-7x+4}$

Rescue Roody!

35. Roody was asked to simplify the following rational expressions:

    (a) $\dfrac{a-4}{4a-16}$

    (b) $\dfrac{y+8}{2y}$

    His work is shown below. All were marked incorrect. Help him understand what went wrong.

    (a) $\dfrac{a-4}{4a-16}=\dfrac{a-4}{4(a-4)}=\dfrac{\cancel{a-4}}{4\cancel{(a-4)}}=4$

    (b) $\dfrac{y+8}{2y}=\dfrac{y+8}{2\cancel{y}}=\dfrac{8}{2}=4$

10.2 Multiplying and Dividing Rational Expressions

Objective: To multiply and divide rational expressions

As we saw in the last section, rational expressions are really just special types of fractions. Thus, to multiply and divide rational expressions, we must use the same techniques as when we multiply and divide fractions. Let us review how to divide and multiply fractions.

Multiplication of Fractions

To multiply two fractions, we use the property $\frac{a}{b} \cdot \frac{c}{d} = \frac{ac}{bd}$. However, we may be able to simplify by dividing common factors first before multiplying, making the process easier. Here are some examples to illustrate.

Example 1 Perform the indicated operation and simplify your answer.

a) $\dfrac{5}{7} \cdot \dfrac{9}{4}$

b) $\dfrac{13}{49} \cdot \dfrac{14}{11}$

Solution.

a) There are no common factors between 5, 4, 7, and 9, so we cannot simplify.

$$\frac{5}{7} \cdot \frac{9}{4} = \frac{5 \cdot 9}{7 \cdot 4} \qquad \text{Multiply straight across}$$

$$= \frac{45}{28} \qquad \text{Our Solution}$$

b) 14 and 49 share a common factor 7. Simplify first before multiplying.

$$\frac{13}{49} \cdot \frac{14}{11} = \frac{13}{\underset{7}{\cancel{49}}} \cdot \frac{\overset{2}{\cancel{14}}}{11} \qquad \text{Divide common factor 7}$$

$$= \frac{13}{7} \cdot \frac{2}{11} \qquad \text{Multiply straight across}$$

$$= \frac{26}{77} \qquad \text{Our Solution}$$

Simplifying first before multiplying made the fraction multiplication much easier. We also arrive at an already simplified fraction for our final result. We will make use of the same idea when we multiply rational expressions.

Division of Fractions

To divide fractions, we first convert the division problem into a multiplication problem. That is, to divide fractions, we multiply the first fraction by the *reciprocal* of the second fraction, as follows:

$$\frac{a}{b} \div \frac{c}{d} = \frac{a}{b} \cdot \frac{d}{c} = \frac{ad}{bc}.$$

Example 2 Perform the indicated operation and simplify your answer.

a) $\dfrac{35}{9} \div 25$

b) $\dfrac{\dfrac{10}{63}}{\dfrac{25}{18}}$

Solution.

a)

$$\frac{35}{9} \div 25 = \frac{35}{9} \div \frac{25}{1}$$ Rewrite 25 as $\dfrac{25}{1}$

$$= \frac{35}{9} \cdot \frac{1}{25}$$ Multiply and reciprocate second fraction

$$= \frac{\overset{7}{\cancel{35}}}{9} \cdot \frac{1}{\underset{5}{\cancel{25}}}$$ Divide common factor 5

$$= \frac{7}{9} \cdot \frac{1}{5}$$ Multiply straight across

$$= \frac{7}{45}$$ Our Solution

b)

$$\frac{\dfrac{10}{63}}{\dfrac{25}{18}} = \frac{10}{63} \div \frac{25}{18}$$ Rewrite given problem

$$= \frac{10}{63} \cdot \frac{18}{25}$$ Multiply and reciprocate second fraction

$$= \frac{\overset{2}{\cancel{10}}}{\underset{7}{\cancel{63}}} \cdot \frac{\overset{2}{\cancel{18}}}{\underset{5}{\cancel{25}}}$$ Divide common factors

$$= \frac{2}{7} \cdot \frac{2}{5}$$ Multiply straight across

$$= \frac{4}{35}$$ Our Solution

Exercise 1 **Class Example**
Perform the indicated operation and simplify your answer.

a) $\dfrac{99}{82} \div \dfrac{41}{66}$

b) $\dfrac{\frac{81}{5}}{9}$

Exercise 2 **You Try**
Perform the indicated operation and simplify your answer.

a) $\dfrac{-4}{7} \div \dfrac{-21}{2}$

b) $144 \div \dfrac{13}{24}$

Multiplication and Division of Rational Expressions

We are now ready to work on rational expressions. We will use the same techniques as in multiplying and dividing fractions. However, it is crucial that the numerator and denominator are **factored** completely before performing the indicated operation.

Example 3 Perform the indicated operation and simplify your answer.

a) $\dfrac{4}{z^2} \cdot \dfrac{z}{2}$

c) $(c^2 - 25) \div \dfrac{5+c}{c-5}$

b) $\dfrac{3y}{2(y+7)^2} \cdot \dfrac{y+7}{6y}$

d) $\dfrac{x}{x^2-1} \div \dfrac{2x^3}{1-x}$

Solution.

a) Simplify first by finding common factors, before multiplying.

$$\frac{4}{z^2} \cdot \frac{z}{2} = \frac{\overset{2}{\cancel{4}}}{\underset{z}{\cancel{z^2}}} \cdot \frac{\overset{1}{\cancel{z}}}{\underset{1}{\cancel{2}}} \qquad \text{Divide common factors}$$

$$\frac{2}{z} \cdot \frac{1}{1} \qquad \text{Multiply straight across}$$

$$= \frac{2}{z} \qquad \text{Our Solution}$$

b) Simplify first by finding common factors, before multiplying.

$$\frac{3y}{2(y+7)^2} \cdot \frac{y+7}{6y} = \frac{\overset{1}{\cancel{3y}}}{\underset{}{2\cancel{(y+7)^2}}} \cdot \frac{\overset{1}{\cancel{(y+7)}}}{\underset{2}{\cancel{6y}}} \qquad \text{Divide common factor}$$

$$= \frac{1}{2(y+7)} \cdot \frac{1}{2} \qquad \text{Multiply straight across}$$

$$= \frac{1}{4(y+7)} \qquad \text{Our Solution}$$

Note. We could have distributed the factor 4 in $\dfrac{1}{4(y+7)}$ giving us $\dfrac{1}{4y+28}$. While both answers are acceptable, we recommend leaving answers in their factored form.

c) Convert the division problem to a multiplication problem and reciprocate the second expression. Recall that $c^2 - 25 = \dfrac{c^2 - 25}{1}$.

$$(c^2 - 25) \div \frac{5+c}{c-5} = \frac{c^2 - 25}{1} \cdot \frac{c-5}{5+c} \qquad \text{Factor } c^2 - 25$$

$$= \frac{(c-5)(c+5)}{1} \cdot \frac{c-5}{5+c} \qquad \text{Recall } c+5 = 5+c$$

$$= \frac{(c-5)\overset{1}{\cancel{(c+5)}}}{1} \cdot \frac{c-5}{\underset{1}{\cancel{(5+c)}}} \qquad \text{Divide common factors}$$

$$= (c-5)^2 \qquad \text{Our Solution}$$

Note. We could have expanded $(c-5)^2$ giving us $c^2 - 10c + 25$. While both answers are acceptable, we recommend leaving answers in their factored form.

d) Convert the division problem to a multiplication problem and reciprocate the second expression.

$$\frac{x}{x^2-2x+1} \div \frac{2x^3}{1-x} = \frac{x}{x^2-2x+1} \cdot \frac{1-x}{2x^3}$$ Factor trinomial

$$= \frac{x}{(x-1)(x-1)} \cdot \frac{-(x-1)}{2x^3}$$ Recall: $1-x=-(x-1)$

$$= \frac{x^{1}}{(x-1)(x-1)} \cdot \frac{-(x-1)^{1}}{2x^{3}_{\ x^2}}$$ Divide common factors

$$= \frac{1}{(x-1)} \cdot \frac{-1}{2x^2}$$ Multiply straight across

$$= \frac{-1}{2x^2(x-1)} \text{ or } -\frac{1}{2x^2(x-1)}$$ Our Solution

Exercise 3 Class Example

Perform the indicated operation and simplify your answer.

a) $\dfrac{10}{n} \cdot \dfrac{3n^2}{5}$

d) $\dfrac{d^2-9}{d+3} \div (d-3)$

b) $(x^2+6x) \cdot \dfrac{x}{6+x}$

e) $\dfrac{y-4}{y} \div \dfrac{4-y}{2y}$

c) $\dfrac{n-7}{n^2-2n-35} \cdot \dfrac{9n+54}{10n+50}$

f) $\dfrac{7+k}{k^2-k-12} \div \dfrac{k^2+7k}{k^2-4k}$

Exercise 4 You Try

Perform the indicated operation and simplify your answer.

a) $\dfrac{9}{4m^2} \cdot \dfrac{m}{3}$

d) $y^2 \div \dfrac{y+4}{y}$

b) $\dfrac{7c}{7+c} \cdot (c^2 - 49)$

e) $\dfrac{6}{3a-1} \div \dfrac{24}{(3a-1)^2}$

c) $\dfrac{5-x}{x^2+3x} \cdot \dfrac{7x+21}{x^2-4x-5}$

f) $\dfrac{2n^2-12n+18}{n+7} \div (2n+6)$

10.2: Exercises

Perform the indicated operation and provide your final answer in simplified form.

1. $\dfrac{6}{7} \cdot \dfrac{5}{48}$

2. $\dfrac{13}{4} \cdot \dfrac{-3}{5}$

3. $\dfrac{55}{14} \cdot \dfrac{28}{15}$

4. $-\dfrac{9}{8} \cdot (-16)$

5. $\dfrac{4}{3} \div \left(-\dfrac{2}{9}\right)$

6. $\dfrac{\frac{14}{5}}{\frac{28}{5}}$

7. $-9 \div \dfrac{-25}{9}$

8. $\dfrac{\frac{36}{7}}{-4}$

Optional Challenge Questions. Perform the indicated operation and provide your final answer in simplified form.

9. $\dfrac{1}{2} \cdot \dfrac{2}{3} \cdot \dfrac{3}{4} \cdot \dfrac{4}{5} \cdot \dfrac{5}{6} \cdot \dfrac{6}{7} \cdot \dfrac{7}{8}$

10. $2 \cdot \dfrac{5}{4} \cdot \dfrac{11}{10} \cdot \dfrac{23}{22} \cdot \dfrac{47}{46}$

Perform the indicated operation and provide your final answer in simplified form.

11. $\dfrac{3}{x^2} \cdot \dfrac{5x}{6}$

12. $\dfrac{2y}{y-1} \cdot \dfrac{5}{4y}$

13. $(2z+6) \div (z^2+3z)$

14. $\dfrac{(n+2)^3}{5n} \cdot \dfrac{10}{n+2}$

15. $\dfrac{4w^2}{4+w} \cdot \dfrac{w^2-16}{w^3}$

16. $\dfrac{a^2+6a+9}{3} \div (3+a)$

17. $\dfrac{y-4}{3y^2} \cdot \dfrac{6y}{4-y}$

18. $\dfrac{-7}{c+5} \div \dfrac{3c}{2c-9}$

19. $\dfrac{6h}{7h+3} \div \dfrac{3}{(7h+3)^2}$

20. $\dfrac{2}{8+c} \cdot \dfrac{c^2-64}{2c-16}$

21. $\dfrac{v-1}{8} \cdot \dfrac{4}{v^2-11v+10}$

22. $\dfrac{g-5}{7-2g} \div \dfrac{5-g}{2g-7}$

23. $\dfrac{p-8}{p^2-12p+32} \div \dfrac{5}{p-4}$

24. $\dfrac{x^2-7x+10}{2-x} \cdot \dfrac{x+4}{x^2-x-20}$

25. $\dfrac{k}{k^2-k-12} \div \dfrac{k^2+4k}{k^2-16}$

Rescue Roody!
Help Roody find the errors and understand what went wrong in the following solutions.

26. Perform the indicated operation and simplify your answer: $\dfrac{x+3}{5} \cdot \dfrac{10}{3+x}$

    Solution:

    $$\frac{x+3}{5} \cdot \frac{10}{3+x} = \frac{-1(3+x)}{5} \cdot \frac{10}{(3+x)}$$

    $$= \frac{-1\cancel{(3+x)}^{1}}{\cancel{5}^{1}} \cdot \frac{\cancel{10}^{2}}{\cancel{(3+x)}^{1}}$$

    $$= \frac{-1}{1} \cdot \frac{2}{1}$$

    $$= -2$$

27. Perform the indicated operation and simplify your answer: $\dfrac{y+8}{y} \cdot \dfrac{y^2}{16}$

    Solution:

    $$\frac{y+8}{y} \cdot \frac{y^2}{16} = \frac{\cancel{y}+8}{\cancel{y}} \cdot \frac{y^2}{16}$$

    $$= \frac{\cancel{8}}{1} \cdot \frac{y^2}{\cancel{16}}$$

    $$= \frac{1}{1} \cdot \frac{y^2}{2}$$

    $$= \frac{y^2}{2}$$

10.3   Adding and Subtracting Rational Expressions

Objective: To add and subtract rational expressions with and without common denominators

Common Denominators

The steps for adding or subtracting rational expressions are the same as when working with fractions. When the denominators are alike, we add or subtract the numerator while keeping the denominator as it is.

Example 1 Perform the indicated operation and simplify your answer.

a) $\dfrac{5}{8} + \dfrac{7}{8}$

b) $\dfrac{5}{2x} - \dfrac{7}{2x}$

Solution.

a) Since the denominators are alike, we add the numerators and keep the denominator.

$$\frac{5}{8} + \frac{7}{8} = \frac{5+7}{8} \qquad \text{Add numerators and keep denominator}$$

$$= \frac{12}{8} \qquad \text{Simplify fraction by dividing common factor 4}$$

$$= \frac{3}{2} \qquad \text{Our Solution}$$

b) Since the denominators are alike, we add the numerators and keep the denominator.

$$\frac{5}{2x} - \frac{7}{2x} = \frac{5-7}{2x} \qquad \text{Subtract numerators and keep denominator}$$

$$= \frac{-2}{2x} \qquad \text{Simplify fraction by dividing common factor 2}$$

$$= \frac{-1}{x} \text{ or } -\frac{1}{x} \qquad \text{Our Solution}$$

Exercise 1 Class Example

Perform the indicated operation and simplify your answer.

a) $\dfrac{5}{3y^2} - \dfrac{14}{3y^2}$

b) $\dfrac{2}{m^2-4} + \dfrac{m}{m^2-4}$

Exercise 2 **You Try**

Perform the indicated operation and simplify your answer.

a) $\dfrac{3}{x+2} + \dfrac{7}{x+2}$

b) $\dfrac{4}{a-2} - \dfrac{2a}{a-2}$

Example 2 Perform the indicated operation and simplify your answer: $\dfrac{2x+5}{x+3} - \dfrac{x-1}{x+3}$

Solution.

Since the denominators are alike, we subtract numerators and keep the denominator. Note that a minus sign is in front of the second fraction and the numerator of the second fraction is not a monomial. In this case, group the terms in each numerator using a parenthesis.

$$\dfrac{2x+5}{x+3} - \dfrac{x-1}{x+3} = \dfrac{(2x+5)-(x-1)}{x+3} \qquad \text{Distribute the negative sign}$$

$$= \dfrac{2x+5-x+1}{x+3} \qquad \text{Perform the indicated operation}$$

$$= \dfrac{x+6}{x+3} \qquad \text{Our Solution}$$

Exercise 3 **Class Example**

Perform the indicated operation and simplify your answer.

a) $\dfrac{2m}{m+5} + \dfrac{10}{m+5}$

c) $\dfrac{x+5}{x-6} - \dfrac{2x-1}{x-6}$

b) $\dfrac{y-3}{y^2-16} + \dfrac{y-5}{y^2-16}$

d) $\dfrac{v^2}{v-5} - \dfrac{10v-25}{v-5}$

Exercise 4 **You Try**

Perform the indicated operation and simplify your answer.

a) $\dfrac{n}{6n+3}+\dfrac{n+1}{6n+3}$

c) $\dfrac{b}{2a-2b}-\dfrac{a}{2a-2b}$

b) $\dfrac{3y}{y+1}-\dfrac{y-2}{y+1}$

d) $\dfrac{8}{n+4}-\dfrac{4n}{n+4}+\dfrac{6n}{n+4}$

Unlike Denominators

What about when the denominators are not the same? We have to first build equivalent fractions with a common denominator. Let us recall how this is done with fractions.

Example 3 Perform the indicated operation and simplify your answer: $\dfrac{3}{10}-\dfrac{5}{12}$

Solution.

Since the denominators are different, we must create equivalent fractions so the original fractions will have a common denominator. To do this, we find the least common denominator (or LCD). We start with the prime factorization of the denominators, 10 and 12.

$$10 = 2 \cdot 5$$
$$12 = 2 \cdot 2 \cdot 3$$

To find the LCD, take all the factors from one denominator, then multiply any missing factors from the other denominator. In this example, take the factors of 10, which are $2 \cdot 5$, and multiply them by the missing factors from 12, which are $2 \cdot 3$. Therefore, the LCD is $2 \cdot 5 \cdot 2 \cdot 3 = 60$.

To create an equivalent fraction for $\dfrac{3}{10}$ with a denominator, 60, we multiply both numerator and denominator by the missing factors, $2 \cdot 3 = 6$, as follows:

$$\frac{3}{10} \cdot \frac{6}{6} = \frac{18}{60}$$

To create an equivalent fraction for $\dfrac{5}{12}$ with a denominator of 60, we multiply both numerator and denominator by the missing factor, 5, as follows:

$$\frac{5}{12} \cdot \frac{5}{5} = \frac{25}{60}$$

Now that we have common denominators, we can subtract the two fractions.

$$\frac{3}{10} - \frac{5}{12} = \frac{18}{60} - \frac{25}{60} \qquad \text{Find equivalent fractions}$$

$$= \frac{18 - 25}{60} \qquad \text{Subtract numerators and keep denominator}$$

$$= \frac{-7}{60} \text{ or } -\frac{7}{60} \qquad \text{Our Solution}$$

Note. Why do we multiply both the numerator and denominator by the missing factors? We want to build equivalent fractions. Meaning, the new fraction has the same value as the original fraction. Multiplying by 1 such as $\frac{6}{6}$ or $\frac{5}{5}$ does not change the value of the original fraction.

Let us now take a look at how to add and subtract rational expressions when we have unlike denominators.

Example 4 Perform the indicated operation and simplify your answer.

a) $\dfrac{4}{x^2} + \dfrac{6}{5x}$

c) $\dfrac{5}{v^2 - v} + \dfrac{6}{v - 1}$

b) $\dfrac{5}{y + 2} - \dfrac{3}{4y}$

d) $\dfrac{8}{z - 3} + \dfrac{2}{3 - z}$

Solution.

a) The denominators are different. We must create equivalent fractions so the original fractions will have a common denominator. To do this, we must find the LCD. Start with the prime factorization of the denominators, x^2 and $5x$.

$$x^2 = x \cdot x$$
$$5x = 5 \cdot x$$

To find the LCD, take the factors of x^2, which are $x \cdot x$, and multiply them by the missing factors from $5x$, which is 5. Therefore, the LCD is $x \cdot x \cdot 5 = 5x^2$.

To create an equivalent fraction for $\dfrac{4}{x^2}$ with a denominator, $5x^2$, we multiply both numerator and denominator by the missing factor, 5, as follows:

$$\frac{4}{x^2} \cdot \frac{5}{5} = \frac{20}{5x^2}$$

To create an equivalent fraction for $\frac{6}{5x}$ with a denominator of $5x^2$, we multiply both numerator and denominator by the missing factor, x, as follows:

$$\frac{6}{5x} \cdot \frac{x}{x} = \frac{6x}{5x^2}$$

Now that we have common denominators, we can add the two fractions.

$$\frac{4}{x^2} + \frac{6}{5x} = \frac{20}{5x^2} + \frac{6x}{5x^2} \qquad \text{Add numerator and keep denominator}$$

$$= \frac{20 + 6x}{5x^2} \text{ or } \frac{6x + 20}{5x^2} \qquad \text{Our Solution}$$

Remark. We can also factor the numerator to get an answer $\frac{2(3x + 10)}{5x^2}$. The expression cannot be further simplified. Either answer is acceptable.

b) The denominators are different. We must create equivalent fractions so the original fractions will have a common denominator. To do this, we must find the LCD. Start with the prime factorization of the denominators, $y + 2$ and $4y$.

$$y + 2 = (y + 2)$$
$$4y = 4 \cdot y$$

Note. $y + 2$ is a binomial that cannot be factored further, y and 2 are *terms* of the binomial, NOT factors.

To find the LCD, take the factor of $(y + 2)$ and multiply it by the missing factors from $4y$, which is $4 \cdot y$. Therefore, the LCD is $(y + 2) \cdot 4 \cdot y = 4y(y + 2)$.

To create an equivalent fraction for $\frac{5}{y + 2}$ with a denominator, $4y(y + 2)$, we multiply both numerator and denominator by the missing factor, $4y$, and to create an equivalent fraction for $\frac{3}{4y}$ with a denominator of $4y(y + 2)$, we multiply both numerator and denominator by the missing factor, $(y + 2)$.

Let us subtract the two fractions.

$$\frac{5}{y+2} - \frac{3}{4y} = \frac{5}{y+2} \cdot \frac{4y}{4y} - \frac{3}{4y} \cdot \frac{(y+2)}{(y+2)} \qquad \text{Find equivalent fractions}$$

$$= \frac{20y}{4y(y+2)} - \frac{3(y+2)}{4y(y+2)} \qquad \text{Subtract fractions}$$

$$= \frac{20y - 3(y+2)}{4y(y+2)} \qquad \text{Distribute } -3$$

$$= \frac{20y - 3y - 6}{4y(y+2)} \qquad \text{Combine like terms}$$

$$= \frac{17y - 6}{4y(y+2)} \qquad \text{Our Solution}$$

c) The denominators are different. We must create equivalent fractions so the original fractions will have a common denominator. To do this, we must find the LCD. Start with the prime factorization of the denominators, $v^2 - v$ and $v - 1$.

$$v^2 - v = v \cdot (v - 1)$$
$$v - 1 = (v - 1)$$

Note. $v - 1$ is a binomial that cannot be factored further. v and 1 are *terms* of the binomial, NOT factors.

To find the LCD, take the factors of $v^2 - v$, which are $v \cdot (v - 1)$, and multiply them by the missing factors from $(v - 1)$, but there are none. Therefore, the LCD is $v \cdot (v - 1)$.

To create an equivalent fraction for $\dfrac{6}{v - 1}$ with a denominator, $v(v - 1)$, we multiply both numerator and denominator by the missing factor, v. Let us add the two fractions.

$$\frac{5}{v^2 - v} + \frac{6}{v - 1} = \frac{5}{v(v - 1)} + \frac{6}{v - 1} \cdot \frac{v}{v} \qquad \text{Find equivalent fractions}$$

$$= \frac{5}{v(v - 1)} + \frac{6v}{v(v - 1)} \qquad \text{Add fractions}$$

$$= \frac{5 + 6v}{v(v - 1)} \text{ or } \frac{6v + 5}{v(v - 1)} \qquad \text{Our Solution}$$

d) The denominators are different. However, $3 - z = -(z - 3)$. The second rational expression becomes:

$$\frac{2}{3 - z} = \frac{2}{-(z - 3)} = -\frac{2}{z - 3}$$

We now have common denominators and can therefore combine the fractions.

$$\frac{8}{z-3} + \frac{2}{3-z} = \frac{8}{z-3} - \frac{2}{z-3}$$ Rewrite $3-z$ as $-(z-3)$

$$= \frac{8-2}{z-3}$$ Subtract numerator

$$= \frac{6}{z-3}$$ Our Solution

Note. The answer cannot be further simplified. The numerator, 6, and denominator, 3, cannot be simplified because 3 is NOT a factor. It is a term in the binomial $(z-3)$.

Exercise 5 **Class Example**

Perform the indicated operation and simplify your answer.

a) $\dfrac{2}{15c} + \dfrac{5}{12c}$

d) $\dfrac{n}{n+3} - \dfrac{9}{n}$

b) $\dfrac{4}{3y} + \dfrac{7}{y^2}$

e) $\dfrac{8}{m^2+2m} - \dfrac{3}{m}$

c) $\dfrac{8}{k-2} - \dfrac{9}{2-k}$

f) $\dfrac{5}{2w} - 3$

Exercise 6 You Try

Perform the indicated operation and simplify your answer.

a) $\dfrac{3}{c} + \dfrac{12}{15c}$

e) $\dfrac{1}{w+2} + \dfrac{1}{w-2}$

b) $\dfrac{5}{y^2} - \dfrac{3}{10y}$

f) $\dfrac{m}{m-5} - \dfrac{2}{m}$

c) $\dfrac{5}{x} + 2$

g) $\dfrac{8}{(p+3)^2} - \dfrac{2}{p+3}$

d) $\dfrac{3}{n} - \dfrac{4}{5n} + \dfrac{7}{2n}$

h) $\dfrac{3}{k} + \dfrac{3}{k^2+6k}$

10.3: Exercises

Perform the indicated operation and simplify your answer.

1. $\dfrac{2}{a+3} + \dfrac{4}{a+3}$

2. $\dfrac{11}{y+2} - \dfrac{6}{y+2}$

3. $\dfrac{7}{3x} + \dfrac{5}{3x}$

4. $\dfrac{3m}{m-5} - \dfrac{15}{m-5}$

5. $\dfrac{6}{r-6} - \dfrac{r}{r-6}$

6. $\dfrac{t^2+4t}{t-1} + \dfrac{2t-7}{t-1}$

7. $\dfrac{x^2}{x-2} - \dfrac{6x-8}{x-2}$

8. $\dfrac{3}{4y-12} - \dfrac{y}{4y-12}$

9. $\dfrac{2w}{5} + \dfrac{7w}{4}$

10. $\dfrac{3}{y} - \dfrac{9}{2y}$

11. $\dfrac{3}{x} + \dfrac{4}{x^2}$

12. $\dfrac{6}{p} - \dfrac{14}{p^2}$

13. $\dfrac{x+5}{8} + \dfrac{x-3}{12}$

14. $\dfrac{a+2}{2} - \dfrac{a-4}{4}$

15. $\dfrac{2}{r} + \dfrac{3}{r-6}$

16. $\dfrac{7}{y+2} - \dfrac{5}{y}$

17. $\dfrac{8}{x-5} + \dfrac{3}{5-x}$

18. $\dfrac{m}{2-m} - \dfrac{4m}{m-2}$

19. $\dfrac{10}{w+2} - 3$

20. $5 + \dfrac{4}{k^2}$

21. $\dfrac{y}{(y+3)^2} + \dfrac{2}{y+3}$

22. $\dfrac{4}{a+1} - \dfrac{4}{(a+1)^2}$

23. $\dfrac{4}{n^2+5n} - \dfrac{1}{n}$

24. $\dfrac{2}{w} + \dfrac{6}{w^2-3w}$

Rescue Roody!

What did Roody do wrong in each of these problems? Identify the error(s), then perform the operation correctly.

25. Perform the indicated operation and simplify the answer: $\dfrac{1}{a} + \dfrac{1}{5}$

Solution.

$$\frac{1}{a}+\frac{1}{5}=\frac{2}{a+5}$$

26. Perform the indeicated operation and simplify the answer: $\dfrac{1}{x}+\dfrac{1}{x+3}$

    Solution. The LCD is $x+3$.

$$\frac{1}{x}+\frac{1}{x+3}=\frac{3}{x+3}+\frac{1}{x+3}$$
$$=\frac{4}{x+3}$$

27. Perform the indicated operation and simplify the answer: $\dfrac{x^2}{x-2}-\dfrac{6x-8}{x-2}$

    Solution.

$$\frac{x^2}{x-2}-\frac{6x-8}{x-2}=\frac{x^2-6x-8}{x-2}$$
$$=\frac{(x-4)(x-2)}{(x-2)}$$
$$=x-4$$

10.4 Complex Fractions

Objective: To simplify complex fractions and complex rational expressions

Complex Fractions

An expression of the form $\dfrac{N}{D}$ is called a **complex fraction**, when the numerator, N, or the denominator, D, or both, contain a fraction. Here are some examples of complex fractions.

$$\frac{\dfrac{4}{3}}{8} \qquad \frac{\dfrac{3}{5}-1}{\dfrac{3}{7}} \qquad \frac{\dfrac{2}{3}}{\dfrac{1}{4}} \qquad \frac{1+\dfrac{6}{5}}{9-\dfrac{4}{3}}$$

Our goal is to a simplify complex fraction into a simple fraction, where the numerator and denominator each contain only a single integer. Two different methods will be illustrated. Method 1 involves combining fractions so that the numerator and denominator each has one single fraction. Then we divide the numerator by the denominator which is the same as multiplying the numerator by the reciprocal of the denominator. Method 2 clears the complex fractions by finding the LCD of all fractions involved and then multiplying the numerator and the denominator by the LCD. With practice, you will develop the skill to decide which method is more efficient to use for each case.

Example 1 Simplify the following complex fractions.

a) $\dfrac{\dfrac{5}{7}}{\dfrac{3}{2}}$
b) $\dfrac{\dfrac{2}{5}}{4}$
c) $\dfrac{\dfrac{5}{9}+\dfrac{1}{3}}{\dfrac{1}{3}}$

Solution.

a) **Method 1** - Both numerator and denominator contain a single fraction.

$$\frac{\dfrac{5}{7}}{\dfrac{3}{2}} = \frac{5}{7} \div \frac{3}{2} \qquad \text{Rewrite as a division problem}$$

$$= \frac{5}{7} \cdot \frac{2}{3} \qquad \text{Multiply and reciprocate second fraction}$$

$$= \frac{10}{12} \qquad \text{Our Solution}$$

Method 2 - For the fractions $\dfrac{5}{7}$ and $\dfrac{2}{3}$, the LCD is 14.

$$\dfrac{\dfrac{5}{7}}{\dfrac{3}{2}} = \dfrac{\dfrac{5}{7}}{\dfrac{3}{2}} \cdot \dfrac{14}{14} \qquad \text{Multiply numerator and denominator by LCD}$$

$$= \dfrac{\left(\dfrac{5}{7}\right) 14}{\left(\dfrac{3}{2}\right) 14} \qquad \text{Simplify}$$

$$= \dfrac{\left(\dfrac{5}{\cancel{7}_1}\right) \cancel{14}^{2}}{\left(\dfrac{3}{\cancel{2}_1}\right) \cancel{14}^{2}} \qquad \text{Divide common factors}$$

$$= \dfrac{10}{21} \qquad \text{Our Solution}$$

b) **Method 1** - Both numerator and denominator contain a single fraction.

$$\dfrac{\dfrac{2}{5}}{4} = \dfrac{\dfrac{2}{5}}{\dfrac{4}{1}} \qquad \text{Multiply and reciprocate denominator}$$

$$= \dfrac{2}{5} \cdot \dfrac{1}{4} \qquad \text{Simplify}$$

$$= \dfrac{\cancel{2}^{1}}{5} \cdot \dfrac{1}{\cancel{4}_{2}} \qquad \text{Divide common factor}$$

$$= \dfrac{1}{10} \qquad \text{Our Solution}$$

Method 2 - There is only one fraction, $\dfrac{2}{5}$ in this problem. Therefore, the LCD is 5.

$$\dfrac{\frac{2}{5}}{4} = \dfrac{\frac{2}{5}}{4} \cdot \dfrac{5}{5} \qquad \text{Multiply numerator and denominator by LCD}$$

$$= \dfrac{\left(\frac{2}{5}\right)(5)}{(4)(5)} \qquad \text{Simplify}$$

$$= \dfrac{\left(\frac{2}{\cancel{5}^{1}}\right)(\cancel{5}^{1})}{(4)(5)} \qquad \text{Simplify}$$

$$= \dfrac{2}{20} \qquad \text{Simplify fraction}$$

$$= \dfrac{1}{10} \qquad \text{Our Solution}$$

c) Note that the $\frac{1}{3}$ in the numerator cannot be canceled with the $\frac{1}{3}$ in the denominator. These are terms and not factors.

Method 1 - Combine the numerator to form a single fraction by first finding the LCD.

$$\dfrac{\frac{5}{9} + \frac{1}{3}}{\frac{1}{3}} = \dfrac{\frac{5}{9} + \frac{1}{3} \cdot \frac{3}{3}}{\frac{1}{3}} \qquad \text{LCD in numerator is 9}$$

$$= \dfrac{\frac{5}{9} + \frac{3}{9}}{\frac{1}{3}} \qquad \text{Add numerator}$$

$$= \dfrac{\frac{8}{9}}{\frac{1}{3}} \qquad \text{Multiply and reciprocate denominator}$$

$$= \dfrac{8}{9} \cdot \dfrac{3}{1} \qquad \text{Simplify}$$

$$= \dfrac{8}{\cancel{9}^{3}} \cdot \dfrac{\cancel{3}^{1}}{1} \qquad \text{Divide common factors}$$

$$= \dfrac{8}{3} \qquad \text{Our Solution}$$

Method 2 - For the fractions $\dfrac{5}{9}$ and $\dfrac{1}{3}$, the LCD is 9.

$$\dfrac{\dfrac{5}{9}+\dfrac{1}{3}}{\dfrac{1}{3}} = \dfrac{\dfrac{5}{9}+\dfrac{1}{3}}{\dfrac{1}{3}}\cdot\dfrac{9}{9}$$ Multiply numerator and denominator by LCD

$$= \dfrac{\left(\dfrac{5}{9}\right)(9)+\left(\dfrac{1}{3}\right)(9)}{\left(\dfrac{1}{3}\right)(9)}$$ Simplify

$$= \dfrac{\left(\dfrac{5}{\cancel{9}_1}\right)(\cancel{9}^1)+\left(\dfrac{1}{\cancel{3}_1}\right)(\cancel{9}^3)}{\left(\dfrac{1}{\cancel{3}_1}\right)(\cancel{9}^3)}$$ Divide common factors

$$= \dfrac{5+3}{3}$$ Add numerator

$$= \dfrac{8}{3}$$ Our Solution

Note. The improper $\frac{8}{3}$ can be written as a mixed number, $2\frac{1}{3}$. Neither one is simpler than the other. There is no need to rewrite the improper fraction as a mixed number.

Exercise 1 **Class Example**
Simplify the following complex fractions.

a) $\dfrac{\dfrac{2}{7}}{\dfrac{4}{5}}$

c) $\dfrac{3 - \dfrac{1}{2}}{\dfrac{3}{5}}$

b) $\dfrac{\dfrac{5}{9}}{8}$

d) $\dfrac{\dfrac{3}{7}}{\dfrac{1}{6} + \dfrac{3}{4}}$

Exercise 2 **You Try**
Simplify the following complex fractions.

a) $\dfrac{\dfrac{2}{5}}{\dfrac{3}{2}}$

c) $\dfrac{\dfrac{3}{4} - \dfrac{2}{3}}{\dfrac{1}{6}}$

b) $\dfrac{2}{\dfrac{1}{4} - 1}$

d) $\dfrac{\dfrac{5}{6} + \dfrac{1}{4}}{\dfrac{5}{2} - 3}$

Complex Rational Expressions

Complex rational expressions or **complex fractional expressions** are expressions that contain one or more rational expressions in the numerator or denominator or both. Here are some examples.

$$\frac{\dfrac{3}{x}}{2-\dfrac{1}{x}} \qquad \frac{8}{\dfrac{4}{y^2}} \qquad \frac{x+\dfrac{1}{2x}}{3-\dfrac{4}{5x}}$$

We will now focus on simplifying complex rational expressions. The same principles in simplifying complex fractions apply in simplifying complex rational expressions. As with the examples above, we will again show the same two methods.

Example 2 Simplify the following complex rational expressions.

a) $\dfrac{\dfrac{4}{x}}{\dfrac{6}{x^2}}$
b) $\dfrac{x+\dfrac{1}{x}}{2}$
c) $\dfrac{3}{1-\dfrac{5}{x}}$

Solution.

a) **Method 1** - Since we have only one rational fraction in the numerator and denominator, we can immediately change the problem to a multiplication problem.

$$\frac{\dfrac{4}{x}}{\dfrac{6}{x^2}} = \frac{4}{x}\cdot\frac{x^2}{6} \qquad \text{Multiply and reciprocate denominator}$$

$$= \frac{\overset{2}{\cancel{4}}}{\cancel{x}^{1}}\cdot\frac{\cancel{x^2}^{\,x}}{\cancel{6}_{3}} \qquad \text{Divide common factors}$$

$$= \frac{2x}{3} \qquad \text{Our Solution}$$

Method 2 - For the rational expressions, $\dfrac{4}{x}$ and $\dfrac{6}{x^2}$, the LCD is x^2.

$$\frac{\frac{4}{x}}{\frac{6}{x^2}} = \frac{\frac{4}{x}}{\frac{6}{x^2}} \cdot \frac{x^2}{x^2}$$ Multiply numerator and denominator by LCD

$$= \frac{\left(\frac{4}{x}\right)(x^2)}{\left(\frac{6}{x^2}\right)(x^2)}$$ Simplify

$$= \frac{\left(\frac{4}{x}\right)(x^2)^x}{\left(\frac{6}{x^2}\right)(x^2)^1}$$ Divide common factors

$$= \frac{4x}{6}$$ Simplify fraction

$$= \frac{2x}{3}$$ Our Solution

b) **Method 1** - We need to find the LCD of the terms in the numerator first so we can add them.

$$\frac{x + \frac{1}{x}}{2} = \frac{\frac{x}{1} \cdot \frac{x}{x} + \frac{1}{x}}{2}$$ LCD of numerator is x

$$= \frac{\frac{x^2}{x} + \frac{1}{x}}{2}$$ Add numerator

$$= \frac{\frac{x^2 + 1}{x}}{2}$$ Multiply and reciprocate denominator

$$= \frac{x^2 + 1}{x} \cdot \frac{1}{2}$$ Recall that $2 = \frac{2}{1}$

$$= \frac{x^2 + 1}{2x}$$ Our Solution

Method 2 - There is only one fraction, $\dfrac{1}{x}$, in the whole expression. So the LCD is x.

$$\frac{x+\dfrac{1}{x}}{2} = \frac{x+\dfrac{1}{x}}{2} \cdot \frac{x}{x}$$ Multiply numerator and denominator by LCD

$$= \frac{\left(x+\dfrac{1}{x}\right)(x)}{(2)(x)}$$ Distribute x in numerator

$$= \frac{(x)(x)+\left(\dfrac{1}{x}\right)(x)}{2x}$$ Simplify

$$= \frac{x^2+\left(\dfrac{1}{x}\right)(x)^1}{2x}$$ Divide common factor

$$= \frac{x^2+1}{2x}$$ Our Solution

c) **Method 1** - We need to find the LCD of the terms in the denominator first so we can subtract them.

$$\frac{3}{1-\dfrac{5}{x}} = \frac{3}{1\cdot\dfrac{x}{x}-\dfrac{5}{x}}$$ LCD of denominator is x

$$= \frac{3}{\dfrac{x}{x}-\dfrac{5}{x}}$$ Subtract denominator

$$= \frac{3}{\dfrac{x-5}{x}}$$ Multiply and reciprocate denominator

$$= \frac{3}{1}\cdot\frac{x}{x-5}$$ Recall $3=\dfrac{3}{1}$; Multiply

$$= \frac{3x}{x-5}$$ Our Solution

Method 2 - There is only one fraction, $\dfrac{5}{x}$, in the whole expression. So the LCD is x.

$$\frac{3}{1-\dfrac{5}{x}} = \frac{3}{1-\dfrac{5}{x}} \cdot \frac{x}{x}$$ Multiply numerator and denominator by LCD

$$= \frac{(3)(x)}{\left(1-\dfrac{5}{x}\right)(x)}$$ Distribute x in denominator

$$= \frac{3x}{(1)(x)-\left(\dfrac{5}{x}\right)(x)}$$ Simplify

$$= \frac{3x}{x-\left(\dfrac{5}{x}\right)(x)^{1}}$$ Divide common factor

$$= \frac{3x}{x-5}$$ Our Solution

Exercise 3 Class Example

Simplify the following complex rational expressions.

a) $\dfrac{\dfrac{4}{3y}}{\dfrac{2}{5y^2}}$

c) $\dfrac{1+\dfrac{3}{x}}{x-\dfrac{1}{4}}$

b) $\dfrac{\dfrac{3}{2x}}{1+\dfrac{1}{x}}$

d) $\dfrac{n-1}{n-\dfrac{1}{4}}$

Exercise 4 **You Try**

Simplify the following complex rational expressions.

a) $\dfrac{\dfrac{8x}{4}}{\dfrac{x}{}}$

c) $\dfrac{1+\dfrac{5}{m}}{m+2}$

b) $\dfrac{\dfrac{3}{2}}{1+\dfrac{1}{3y}}$

d) $\dfrac{1+\dfrac{1}{x}}{1-\dfrac{1}{x}}$

10.4: Exercises

Simplify the complex fraction.

1. $\dfrac{\dfrac{1}{2x}}{\dfrac{3}{4x^2}}$

2. $\dfrac{\dfrac{x}{5}}{\dfrac{2}{x^2}}$

3. $\dfrac{\dfrac{3}{7x}}{\dfrac{6x}{5}}$

4. $\dfrac{\dfrac{4x}{9}}{8x}$

5. $\dfrac{\dfrac{2}{x}+1}{\dfrac{5}{6}}$

6. $\dfrac{1-\dfrac{x}{8}}{4}$

7. $\dfrac{4x-\dfrac{2}{5}}{\dfrac{2}{3}}$

8. $\dfrac{3x}{\dfrac{6x}{5}-1}$

9. $\dfrac{2}{\dfrac{1}{x}-\dfrac{1}{9}}$

10. $\dfrac{\dfrac{x}{2}}{x+\dfrac{3}{5}}$

11. $\dfrac{\dfrac{7x}{2}}{5-\dfrac{3}{2x}}$

12. $\dfrac{1+\dfrac{2}{x}}{\dfrac{2}{3x}+1}$

13. $\dfrac{2-\dfrac{2}{x}}{3+\dfrac{1}{x^2}}$

14. $\dfrac{1}{1-\dfrac{1}{x}}$

15. $\dfrac{x+6}{\dfrac{1}{3}+\dfrac{1}{x}}$

16. $\dfrac{\dfrac{2}{x}}{\dfrac{1}{x}-5x}$

17. $\dfrac{\dfrac{3}{7}-x}{\dfrac{3}{x}+2}$

18. $\dfrac{\dfrac{y}{9}-\dfrac{3}{y}}{\dfrac{9}{y}}$

19. $\dfrac{\dfrac{1}{a}+\dfrac{1}{8}}{\dfrac{1}{a}-\dfrac{1}{8}}$

20. $\dfrac{1+\dfrac{2}{c}}{1-\dfrac{4}{c^2}}$

10.5   Solving Rational Equations

Objective: To solve rational equations and proportions

When solving rational equations, we will clear the fraction by multiplying each fraction by the lowest common denominator (or LCD). This is basically the same strategy as solving linear equations with fractions. However, with solving rational equations, since some expression may contain a variable in the denominator, we have to be careful that our solution does not make any rational expression undefined. The following examples show the difference and similarity between how a linear equation and a rational equation look and how each is solved.

Example 1 Solve $\frac{2}{3}x + \frac{7}{6} = \frac{3}{4}x$ for x.

Solution.

This is a linear equation. Note the position of the variable, x. Each fraction will never be undefined. Therefore, no values have to be excluded. LCD for the equation is 12.

$$\frac{2}{3}x + \frac{7}{6} = \frac{3}{4}x \qquad\qquad \text{Multiply each side by the LCD, 12}$$

$$12 \cdot \left(\frac{2}{3}x + \frac{7}{6}\right) = 12 \cdot \left(\frac{3}{4}x\right) \qquad\qquad \text{Distribute 12}$$

$$12 \cdot \left(\frac{2}{3}x\right) + 12 \cdot \left(\frac{7}{6}\right) = 12 \cdot \left(\frac{3}{4}x\right) \qquad\qquad \text{Simplify each term}$$

$$\cancel{12}^{4} \cdot \left(\frac{2}{\cancel{3}_{1}}x\right) + \cancel{12}^{2} \cdot \left(\frac{7}{\cancel{6}_{1}}\right) = \cancel{12}^{3} \cdot \left(\frac{3}{\cancel{4}_{1}}x\right) \qquad\qquad \text{Divide common factors}$$

$$8x + 14 = 9x \qquad\qquad \text{Subtract 8x from each side}$$

$$14 = x \qquad\qquad \text{Our Solution}$$

Verify that we have the correct solution.

$$\frac{2}{3}x + \frac{7}{6} = \frac{3}{4}x$$ Substitute $x = 14$

$$\frac{2}{3}(14) + \frac{7}{6} \overset{?}{=} \frac{3}{4}(14)$$ Simplify fraction on the right

$$\frac{2}{3}(14) + \frac{7}{6} \overset{?}{=} \frac{3}{\underset{2}{\cancel{4}}}(\cancel{14})^{7}$$ Divide common factors

$$\frac{28}{3} + \frac{7}{6} \overset{?}{=} \frac{21}{2}$$ Find LCD for fractions on the left

$$\frac{56}{6} + \frac{7}{6} \overset{?}{=} \frac{21}{2}$$ Add fractions on the left

$$\frac{63}{6} = \frac{21}{2} \quad \checkmark$$ Therefore, our solution is $x = 14$.

Example 2 Solve $\dfrac{2}{3x} + \dfrac{7}{6} = \dfrac{3}{x}$ for x.

Solution.

This is a rational equation. Note the position of the variable, x. The rational expressions $\frac{2}{3x}$ and $\frac{3}{x}$ become undefined when $x = 0$. Therefore, $x = 0$ has to be excluded from the solution. LCD for the equation is 6x.

$$\frac{2}{3x} + \frac{7}{6} = \frac{3}{x}$$ Multiply each side by the LCD, 6x

$$6x \cdot \left(\frac{2}{3x} + \frac{7}{6}\right) = 6x \cdot \left(\frac{3}{x}\right)$$ Distribute 6x on the left

$$6x \cdot \left(\frac{2}{3x}\right) + 6x \cdot \left(\frac{7}{6}\right) = 6x \cdot \left(\frac{3}{x}\right)$$ Simplify each term

$$\cancel{6x}^{2}\left(\frac{2}{\cancel{3x}^{1}}\right) + \cancel{6}x \cdot \left(\frac{7}{\cancel{6}_{1}}\right) = 6\cancel{x}^{1} \cdot \left(\frac{3}{\cancel{x}^{1}}\right)$$ Divide common factors

$$4 + 7x = 18$$ Subtract 4 from each side

$$7x = 14$$ Divide each side by 7

$$x = 2$$ Our Solution

Verify that we have the correct solution.

$$\frac{2}{3x} + \frac{7}{6} = \frac{3}{x}$$ Substitute $x = 2$

$$\frac{2}{3(2)} + \frac{7}{6} \stackrel{?}{=} \frac{3}{(2)}$$ Multiply denominator of first fraction

$$\frac{2}{6} + \frac{7}{6} \stackrel{?}{=} \frac{3}{2}$$ Add fractions on the left

$$\frac{9}{6} = \frac{3}{2} \quad \checkmark$$

Therefore, our solution is $x = 2$.

Exercise 1 **Class Example**

Identify if the equation is linear or rational. Then solve for x.

a) $\dfrac{x}{5} - \dfrac{7}{15} = \dfrac{2}{3}x$

b) $\dfrac{4}{x} + 3 = \dfrac{5}{2x}$

Exercise 2 You Try

Identify if the equation is linear or rational. Then solve for x.

a) $\dfrac{13}{8}x - \dfrac{1}{4} = \dfrac{1}{16} + x$

b) $\dfrac{1}{x} - \dfrac{1}{2} = \dfrac{5}{4x}$

Strategy for Solving Rational Equations:

1. Find the lowest common denominator (or LCD).

2. Identify the values which make each term undefined. These values have to be excluded from the solution.

3. Multiply each term by the LCD.

4. Solve the resulting equation.

5. Verify that each solution is correct.

6. State your solution(s).

Example 3 Solve $\dfrac{5}{2p-1} = \dfrac{3}{p}$ for p.

Solution.

The LCD for the equation is $p(2p-1)$. The rational expression, $\dfrac{5}{2p-1}$, becomes undefined when $p = \dfrac{1}{2}$ and the expression, $\dfrac{3}{p}$, becomes undefined when $p = 0$. Therefore, $p = \dfrac{1}{2}$ and

$p = 0$ have to be excluded from the solution.

$$\frac{5}{2p-1} = \frac{3}{p}$$ Multiply each side by the LCD, $p(2p-1)$

$$p(2p-1) \cdot \left(\frac{5}{2p-1}\right) = p(2p-1) \cdot \left(\frac{3}{p}\right)$$ Simplify each term

$$p\cancel{(2p-1)}^1 \cdot \left(\frac{5}{\cancel{2p-1}}\right) = \cancel{p}(2p-1) \cdot \left(\frac{3}{\cancel{p}^1}\right)$$ Divide common factors

$$p(5) = (2p-1)(3)$$ Multiply
$$5p = 6p - 3$$ Subtract 6p from each side
$$-p = -3$$ Multiply each term by -1
$$p = 3$$ Our Solution

Verify that we have the correct solution.

$$\frac{5}{2p-1} = \frac{3}{p}$$ Substitute $p = 3$

$$\frac{5}{2(3)-1} \overset{?}{=} \frac{3}{(3)}$$ Multiply denominator of first fraction

$$\frac{5}{6-1} \overset{?}{=} \frac{3}{3}$$ Simplify

$$\frac{5}{5} = 1 \ \checkmark$$

Therefore, our solution is $p = 3$.

Example 4 Solve $x + 5 = \dfrac{6}{x}$ for x.

Solution.

The LCD for the equation is x. The rational expression, $\dfrac{6}{x}$, becomes undefined when $x = 0$. Therefore, $x = 0$ has to be excluded from the solution.

$$x + 5 = \frac{6}{x}$$
Multiply each side by the LCD, x

$$x \cdot (x + 5) = x \cdot \left(\frac{6}{x}\right)$$
Distribute x on the left

$$x(x) + x(5) = x\left(\frac{6}{x}\right)$$
Perform indicated operation

$$x^2 + 5x = \cancel{x}^1\left(\frac{6}{\cancel{x}^1}\right)$$
Divide common factors

$$x^2 + 5x = 6$$
Solve quadratic equation

$$x^2 + 5x - 6 = 0$$
Factor trinomial

$$(x + 6)(x - 1) = 0$$
Use zero-product property

$$x + 6 = 0 \text{ or } x - 1 = 0$$
Solve for x

$$x = 6 \text{ or } x = 1$$
Our Solution

Verify that each solution is correct by substituting each solution into the original equation.

Check $x = -6$:

$$x + 5 = \frac{6}{x}$$

$$(-6) + 5 \overset{?}{=} \frac{6}{(-6)}$$

$$-1 = -1 \ \checkmark$$

Check $x = 1$:

$$x + 5 = \frac{6}{x}$$

$$(1) + 5 \overset{?}{=} \frac{6}{(1)}$$

$$6 = 6 \ \checkmark$$

Therefore our solution is $x = -6$ or $x = 1$.

Example 5 Solve $3x - \dfrac{2x}{x+2} = \dfrac{x^2}{x+2}$ for x.

Solution.

The LCD for the equation is $(x+2)$. The rational expressions, $\dfrac{2x}{x+2}$ and $\dfrac{x^2}{x+2}$, become unde-fined when $x = -2$. Therefore, $x = -2$ has to be excluded from the solution.

$$3x - \frac{2x}{x+2} = \frac{x^2}{x+2}$$ Multiply each side by LCD

$$(x+2) \cdot \left(3x - \frac{2x}{x+2}\right) = (x+2) \cdot \left(\frac{x^2}{x+2}\right)$$ Distribute $(x+2)$ on the left

$$(x+2)(3x) - (x+2)\left(\frac{2x}{x+2}\right) = (x+2)\left(\frac{x^2}{x+2}\right)$$ Perform indicated operation

$$(x+2)(3x) - \cancel{(x+2)}^{1}\left(\frac{2x}{\cancel{x+2}}\right) = \cancel{(x+2)}^{1}\left(\frac{x^2}{\cancel{x+2}}\right)$$ Divide common factors

$$3x^2 + 6x - 2x = x^2$$ Combine like terms

$$3x^2 + 4x = x^2$$ Solve quadratic equation

$$2x^2 + 4x = 0$$ Factor binomial

$$2x(x+2) = 0$$ Use zero-product property

$$2x = 0 \text{ or } x+2 = 0$$ Solve for x

$$x = 0 \text{ or } x = -2$$ Our Solution

$x = -2$ is an extraneous solution since $x = -2$ makes the rational expressions, $\dfrac{2x}{x+2}$ and $\dfrac{x^2}{x+2}$ undefined. Verify that $x = 0$ is the correct solution.

$$3x - \frac{2x}{x+2} = \frac{x^2}{x+2}$$ Substitute $x = 0$

$$3(0) - \frac{2(0)}{(0)+2} \stackrel{?}{=} \frac{(0)^2}{(0)+2}$$ Perform indicated operation

$$0 - \frac{0}{2} \stackrel{?}{=} \frac{0}{2}$$ Simplify fractions

$$0 - 0 = 0 \checkmark$$

Therefore, $x = 0$ is our only solution.

Example 6 Solve $\dfrac{2y}{y+1} = \dfrac{3}{y-1}$ for y.

Solution.

The LCD for the equation is $(y+1)(y-1)$. The rational expression, $\dfrac{2y}{y+1}$, becomes undefined when $y=-1$ and the rational expression, $\dfrac{3}{y-1}$, becomes undefined when $y=1$. Therefore, $y=1$ and $y=-1$ have to be excluded from the solution.

$$\frac{2y}{y+1} = \frac{3}{y-1} \qquad \text{Multiply by the LCD}$$

$$(y+1)(y-1)\cdot\left(\frac{2y}{y+1}\right) = (y+1)(y-1)\cdot\left(\frac{3}{y-1}\right) \qquad \text{Simplify each term}$$

$$\cancel{(y+1)}(y-1)\cdot\left(\frac{2y}{\cancel{y+1}}\right) = (y+1)\cancel{(y-1)}\cdot\left(\frac{3}{\cancel{y-1}}\right) \qquad \text{Divide common factors}$$

$$(y-1)(2y) = (y+1)(3) \qquad \text{Distribute}$$

$$2y^2 - 2y = 3y + 3 \qquad \text{Solve the quadratic equation}$$

$$2y^2 - 5y - 3 = 0 \qquad \text{Factor the trinomial}$$

$$(2y+1)(y-3) = 0 \qquad \text{Use zero-product property}$$

$$2y + 1 = 0 \text{ or } y - 3 = 0 \qquad \text{Solve for } y$$

$$y = -\frac{1}{2} \text{ or } y = 3 \qquad \text{Our Solution}$$

Verify that each solution is correct by substituting each solution into the original equation.

Check $y = -\dfrac{1}{2}$:

$$\frac{2y}{y+1} = \frac{3}{y-1}$$

$$\frac{2\left(-\frac{1}{2}\right)}{-\frac{1}{2}+1} \overset{?}{=} \frac{2}{\left(-\frac{1}{2}\right)-1}$$

$$\frac{-1}{\frac{1}{2}} \overset{?}{=} \frac{3}{-\frac{3}{2}}$$

$$-2 = -2 \checkmark$$

Check $y = 3$:

$$\frac{2y}{y+1} = \frac{3}{y-1}$$

$$\frac{2(3)}{(3)+1} \overset{?}{=} \frac{3}{(3)-1}$$

$$\frac{6}{4} = \frac{3}{2} \checkmark$$

Therefore, our solution is $y = -\dfrac{1}{2}$ or $y = 3$.

Exercise 3 Class Example

Solve the following equations.

a) $\dfrac{8}{n+4} = \dfrac{6}{n-1}$

d) $\dfrac{x^2}{x-1} = 2x + \dfrac{1}{x-1}$

b) $3y = \dfrac{1}{y} - \dfrac{1}{2}$

e) $\dfrac{1}{p^2} + \dfrac{1}{p} = 2$

c) $\dfrac{2m}{m-3} = \dfrac{m}{m+1}$

f) $\dfrac{n}{n-3} - \dfrac{4}{5(n-3)} = \dfrac{1}{2}$

Exercise 4 You Try

Solve the following equations.

a) $\dfrac{2}{p-3} = \dfrac{5}{p}$

d) $m + \dfrac{8}{m-4} = \dfrac{5m}{m-4}$

b) $\dfrac{x-3}{x} = 5 - \dfrac{3}{x}$

e) $3 + \dfrac{1}{y} = \dfrac{2}{y^2}$

c) $\dfrac{2n-1}{n} = \dfrac{2n}{n+1}$

f) $\dfrac{5}{2(a+1)} - 3 = \dfrac{2a}{a+1}$

10.5: Exercises

Solve the following equations.

1. $\dfrac{5}{3} = \dfrac{1}{2} + \dfrac{7}{x}$

2. $\dfrac{4}{m} = \dfrac{8}{m-1}$

3. $\dfrac{6}{y+2} = 9$

4. $5 - \dfrac{11}{3z} = \dfrac{2}{3}$

5. $\dfrac{3}{n} = \dfrac{6}{2n-1}$

6. $\dfrac{y}{5(y-1)} + \dfrac{3}{5} = \dfrac{1}{y-1}$

7. $\dfrac{x-3}{2x} = 5 - \dfrac{3}{x}$

8. $\dfrac{4}{p-2} = \dfrac{2p}{p-2} - 1$

9. $\dfrac{c}{9} = \dfrac{4}{c}$

10. $\dfrac{1}{y} + 2y = 3$

11. $\dfrac{x}{3} = \dfrac{4}{x-1}$

12. $a - \dfrac{12}{a} = 4$

13. $\dfrac{n}{n+1} = \dfrac{1}{2n}$

14. $\dfrac{2w}{w-3} - w = \dfrac{6}{w-3}$

15. $\dfrac{p-4}{p-1} = \dfrac{12}{3-p}$

16. $x = \dfrac{10}{x+1} + 8$

17. $\dfrac{2}{3(a-1)} - 1 = \dfrac{4a}{a-1}$

18. $\dfrac{3}{c^2} + \dfrac{1}{c} = 2$

Answer Key

6.1: Answers

1. $5(9x^2 - 5)$

2. $7(8 - 5p)$

3. $10(5x - 8y)$

4. $7ab(1 - 5a)$

5. $-3a^2b(1 - 2ab)$

6. $-8n(4n^2 - 4n - 1)$

7. $-5x^2(1 - 5x + 3x^2)$

8. $3(7p^2 + 10p + 9)$

9. $-2x(5x^3 - 10x - 6)$

10. $5(6b^2 + ab - 3a^2)$

11. $-3y^2(9x^2 - 4x + 3)$

12. $(3x^2 + 2)(x + 5)$

13. $(n^2 - 3)(2n - 2)$

14. $(8r^2 - 5)(5r - 1)$

15. $(3b^2 - 7)(5b + 7)$

16. $(7x + 5)(y - 7)$

17. $(8x + 1)(2y - 7)$

18. $(7n^2 - 5)(n + 3)$

19. $(m + 5)(3n - 8)$

20. $(8x - 1)(y + 7)$

21. $(7p^2 + 5)(4p + 3)$

22. $(2v^2 - 1)(7v + 5)$

23. $(4(y^2 + 2)(y - 3)$

24. $2(5x^2 - 1)(3x - 2)$

25. $2(u + 3)(7u + 2v)$

6.2: Answers

1. $(p+6)(p+6)$

2. $(x+3)(x-4)$

3. $(n-8)(n-1)$

4. $(x-5)(x+6)$

5. $(x+1)(x-10)$

6. $(b+8)(b+4)$

7. Prime

8. $(b-10)(b-7)$

9. $(x-7)(x+10)$

10. $(x-3)(x+6)$

11. $(n-5)(n-3)$

12. $(a+3)(a-9)$

13. $(p+6)(p+9)$

14. $(p+10)(p-3)$

15. $(n-8)(n-7)$

16. $(x+3)(x-3)$

17. $(m+7)(m-7)$

18. $(p+1)(p-1)$

19. $(x-1)^2$

20. $(m+6)^2.$

21. $(y+8)^2$

22. Prime

23. $3(v^2-4v+6)$

24. $-(x-9)^2$

25. $6(x+2)(x+1)$

26. $4(x+3)(x+2)$

27. $-5(n-8)(n-1)$

28. $5v(v-1)(v+5)$

29. $4(n+4)(n-4)$

30. $y(y+5)(y-5)$

6.3: Answers

1. $(2x+3)(2x-1)$

2. $(5x+2)(x+6)$

3. Prime

4. $(3x-1)(2x+3)$

5. $(7x+3)(x-2)$

6. $(3x+2)(3x-4)$

7. $(2x-1)(4x+3)$

8. $(5x+2)(2x+1)$

9. $(5x+3)(3x+1)$

10. $(2x+1)(x+12)$

11. $2(3x-2)(2x+1)$

12. $(7x+3)(3x-2)$

13. $(7x+2)(4x-3)$

14. $(4x-3)(2x+5)$

15. $(5x-2)(4x+3)$

6.4: Answers

1. $(x+3)(x-3)$

2. $(x+1)(x-1)$

3. Prime

4. $(2+p)(2-p)$

5. $(2v+1)(2v-1)$

6. $(3k+2)(3k-2)$

7. $(1+3a)(1-3a)$

8. $3(x+3)(x-3)$

9. $5(25x^2+9)$

10. $5(n+2)(n-2)$

11. $2(3a+5)(3a-5)$

12. $4(16+m^2)$

13. $(a-1)^2$

14. $(k+2)^2$

15. $(n-4)^2$

16. $(5p-1)^2$

17. $(2k+7)^2$

18. $(x+4)^2$

19. $(5a+3)^2$

20. $(2a-5)^2$

21. $2(3m-2)^2$

22. $5(x+1)^2$

23. $5(2x+1)^2$

24. $2(2x-3)^2$

25. $(a^2+9)(a+3)(a-3)$

26. $(n^2+1)(n+1)(n-1)$

27. $(4+z^2)(2+z)(2-z)$

28. Roody can still factor out some common factors.

6.5: Answers

1. $(3p+4)(2p+1)$

2. $(x+3)(x+1)$

3. $(m+2)(m-2)$

4. $v(v^2+1)$

5. Prime

6. $(2x-5)(x-3)$

7. $2(x-2)(x-3)$

8. $(b-3)(a+2)$

9. $n(n+1)(n-6)$

10. $(x+4)^2$

11. $(5x+3)(x-5)$

12. $n(n+2)(n+5)$

13. $(2y^2+5)(4y+3)$

14. $(4a+3)(4a-3)$

15. $(5n-3)(n+2)$

16. $(4x-9)^2$

17. $(2k+5)(k-2)$

18. $3n(3n^2-2n+1)$

19. $2x(x+5)(x-2)$

20. $(x-5)(3y-1)$

21. $(3+5y)(3-5y)$

22. $4(x+3y)^2$

23. $3k(k-5)(k-4)$

24. $(5u-4)(u-1)$

25. $5(3m-5)^2$

26. $3(3m+4)(3m-4)$

27. $x^2(x^2+4)$

28. $n(1+n)(1-n)$

29. $(p^2+9)(p+3)(p-3)$

30. $(y+5)(y+2)(y-2)$

6.6: Answers

1. $x=-3,-2$

2. $x=0,6$

3. $p=-3,0$

4. $t=0,1$

5. $x=-\frac{6}{5},0$

6. $m=0,2$

7. $x=-3,-1$

8. $m=-2,4$

9. $p=4,6$

10. $n=0,6$

11. $x=3,4$

12. $x=-\frac{3}{2},2$

13. $a=-3,2$

14. $n=3,12$

15. $x=10,13$

16. $y=4,11$

17. $n=-10,-3$

18. $p=-1,0,1$

19. $x=-\frac{1}{2}$

20. $n=-6,-1$

21. $x=\frac{1}{2},5$

22. $k=\frac{1}{3},\frac{5}{2}$

23. $n=-\frac{1}{3}$

24. $x=-6,0$

25. $x=-10,25$

26. $x^2-6x+5=0$

27. $x^2+2x-15=0$

28. $x^2-16=0$

29. $x^2-4x-21=0$

30. $x^2-6x+9=0$

31. $x^2-8x+16=0$

32. $3x^2-14x+8=0$

33. $6x^2-7x-3=0$

34. $10x^2-51x+5=0$

35. $25x^2+40x+16=0$

7.1: Answers

1. 10

2. −8

3. 6

4. −11

5. 2

6. Not Real

7. 5

8. 4

9. $\frac{5}{4}$

10. 2

11. $\frac{9}{2}$

12. $-\frac{2}{3}$

13. 3.9

14. Not Real

15. -4.6

16. 10.6

17. -1.2

18. -1.5

19. 18.7

20. 3.4

21. 0.7

7.2: Answers

1. $2\sqrt{2} \approx 2.83$

2. $4\sqrt{2} \approx 5.66$

3. $-3\sqrt{5} \approx -6.71$

4. Not Real

5. $\frac{3\sqrt{2}}{2} \approx 2.12$

6. $\frac{2\sqrt{5}}{3} \approx 1.49$

7. $3 + 5\sqrt{2} \approx 10.07$

8. $7 - 2\sqrt{3} \approx 3.54$

9. $\frac{1}{2} + \frac{3\sqrt{7}}{2} \approx 4.47$

10. $3 + \sqrt{7} \approx 5.65$

11. $\frac{5 + 2\sqrt{6}}{5}$ or

    $1 + \frac{2\sqrt{6}}{5} \approx 1.98$

12. $\frac{5 - \sqrt{10}}{2} \approx 0.92$

s

8.1: Answers

1. $\left| \right|$; Square Root

2. 7

3. 7

4. $|A|$

5. $|A + B|$

6. $|x + 3|$

7. $|2y - 9|$

8. $|x + 3|$

9. $x = \pm 6$

10. $y = \pm\sqrt{11} \approx \pm 3.32$

11. $z = \pm 4\sqrt{2} \approx \pm 5.66$

12. Solutions are complex

13. $x = \pm\sqrt{13} \approx \pm 3.61$

14. $x = \pm\sqrt{5} \approx \pm 2.24$

15. $y = \pm\sqrt{6} \approx \pm 2.45$

16. $y = \pm 6\sqrt{2} \approx \pm 8.49$

17. $z = 2\sqrt{2} \approx \pm 2.83$

18. $v = \pm 10\sqrt{3} \approx \pm 17.32$

19. $t = \pm\frac{1}{3}$

20. $x = 3 + \sqrt{7} \approx 5.65$ or

    $x = 3 - \sqrt{7} \approx 0.35$

21. $z = 1 + \sqrt{2} \approx 2.41$ or

    $z = 1 - \sqrt{2} \approx -0.41$

22. $y = 3$ or $y = -11$

23. $h = 0$ or $h = -4$

24. $t = \frac{-5+\sqrt{10}}{2} \approx -0.92$ or

$t = \frac{-5-\sqrt{10}}{2} \approx -4.08$

25. $y = \frac{-4+\sqrt{3}}{2} \approx -1.13$ or

$y = \frac{-4-\sqrt{3}}{2} \approx -2.87$

26. $g = 2+\sqrt{2} \approx 3.41$ or

$g = 2-\sqrt{2} \approx 0.59$

27. $p = 1$ or $p = -2$

28. $t = \pm 2\sqrt{6} \approx \pm 4.90$

29. Solutions are complex

30. $x = 8+\sqrt{5} \approx 10.24$ or

$x = 8-\sqrt{5} \approx 5.76$

31. $y = \frac{-4+2\sqrt{2}}{3} \approx -0.39$ or

$y = \frac{-4-2\sqrt{2}}{3} \approx -2.28$

8.2: Answers

1. $4; (a+2)^2$

2. $36; (x-6)^2$

3. $9; (m-3)^2$

4. $25; (n+5)^2$

5. $\frac{1}{4}; (y-\frac{1}{2})^2$

6. $\frac{25}{4}; (x+\frac{5}{2})^2$

7. $\frac{1}{36}; (r-\frac{1}{6})^2$

8. $\frac{4}{25}; (p+\frac{2}{5})^2$

9. $n = 1$ or $n = 7$

10. $y = 2$ or $y = -6$

11. $g = 2$ or $g = -4$

12. $x = 4+2\sqrt{7}$ or $x = 4-2\sqrt{7}$;

$x \approx 9.29$ or $x \approx -1.29$

13. No Real Solutions

14. $b = \frac{3}{2}$ or $b = -\frac{7}{2}$

15. $m = \frac{9+\sqrt{21}}{2}$ or $m = \frac{9-\sqrt{21}}{2}$;

$m \approx 6.79$ or $m \approx 2.21$

16. $v = 7+\sqrt{85}$ or $v = 7-\sqrt{85}$

$v \approx 16.22$ or $v \approx -2.22$

17. $k = 1-\sqrt{10}$ or $k = 1+\sqrt{10}$;

$k \approx -2.16$ or $k \approx 4.16$

18. $x = -8+\sqrt{14}$ or $x = -8-\sqrt{14}$;

$x \approx -4.26$ or $x \approx -11.74$

19. $k = 1+\sqrt{5}$ or $k = 1-\sqrt{5}$;

$k \approx 3.24$ or $k \approx -1.24$

20. $b = \frac{-7+\sqrt{61}}{2}$ or $b = \frac{-7-\sqrt{61}}{2}$;

$b \approx 0.41$ or $b \approx -14.81$

21. $a = \frac{-4+\sqrt{17}}{2}$ or $a = \frac{-4-\sqrt{17}}{2}$;

$a \approx 0.06$ or $a \approx -4.06$

22. $y = \frac{4+\sqrt{41}}{5}$ or $y = \frac{4-\sqrt{41}}{5}$;

$y \approx 2.08$ or $y \approx -0.48$

23. $p = \frac{1+\sqrt{19}}{3}$ or $p = \frac{1-\sqrt{19}}{3}$;

$p \approx 1.79$ or $p \approx -1.12$

24. $w = \frac{2}{3}$ or $w = -1$

8.3: Answers

1. $y = \frac{-3\pm\sqrt{13}}{2}$; $y \approx 0.30$ or -3.30

2. $v = 1$ or $v = 3$

3. $p = \frac{-5\pm\sqrt{37}}{2}$; $p \approx 0.54$ or -5.54

4. $m = \pm\frac{\sqrt{6}}{2}$; $m \approx \pm1.22$

5. $g = \frac{2}{3}$

6. $y = 2\pm\sqrt{5}$; $y \approx 4.24$ or -0.24

7. $r = 1$ or $r = -\frac{1}{3}$

8. $x = \frac{1\pm\sqrt{31}}{2}$; $x \approx 3.28$ or -2.28

9. No Real Solutions

10. $k = \frac{3\pm\sqrt{29}}{2}$; $k \approx 4.19$ or -1.19

11. $n = \frac{7}{2}$ or $n = -7$

12. $b = -3$ or $b = \frac{1}{2}$

13. $n = \pm\frac{\sqrt{30}}{3}$; $n \approx \pm1.83$

14. $r = -3\pm\sqrt{5}$;

    $r \approx -0.76$ or -5.24

15. $y = -\frac{1}{2}$

16. $v = \frac{3\pm\sqrt{33}}{6}$; $v \approx 1.46$ or -0.46

17. $x = -1$ or $x = -\frac{3}{2}$

18. $n = -\frac{3}{4}$

8.4: Answers

1. (a) $x = 2, 4$; Factoring is easiest.

   (b) $y = 0, 12$; Factoring is easiest.

   (c) $b = \pm9$; Square Root Property / Completing the Square is easiest.

   (d) $z = -2\pm\sqrt{7}$; either the Square Root Property / Completing the Square or the Quadratic Formula is easiest; Factoring is not an option.

   (e) $x = 6\pm2\sqrt{3}$; Square Root Property / Completing the Square is easiest.

   (f) $v = \frac{3}{2}, -4$; Factoring is easiest.

2. (a) Solve by factoring. (Solution: $y = -\frac{7}{3}, 5$)

   (b) Write the equation in standard form and then solve by factoring. (Solution: $y = -\frac{7}{3}, 5$)

   (c) Divide each side of the equation by 3 to get $z^2 = 4$. Then use the Square Root Property to solve. (Solution: $z = \pm2$)

   (d) Divide each side by 3, then subtract $4z$ from each side and then solve the equation by factoring. (Solution: $z = 0, 4$)

   (e) Write the equation in standard form then use the quadratic formula, or complete the square and then use the square root property. (Solution: $A = -5\pm3\sqrt{3}$)

   (f) Isolate the square term $(2b-7)^2$ and then use the square root property. (Solution: $b = \frac{7\pm\sqrt{5}}{3}$)

   (g) Since there is a square term $(2b-7)^2$ and a linear term, $10b$, first expand the square term and simplify to put the equation in standard form: $4b^2 - 38b + 54 = 0$. Then use the quadratic formula. (Solution: $b = \frac{19\pm\sqrt{145}}{4}$)

3.

(a) $x = 0, 3$

(b) $x = -1, 4$

(c) $y = 0, 3$

(d) $y = \pm\sqrt{3}$

(e) $x = -\frac{3}{2}, \frac{1}{2}$

(f) $x = \frac{7 \pm \sqrt{5}}{2}$

(g) $x = \frac{7}{2}$

(h) $v = -\frac{4}{3}, \frac{3}{2}$

(i) $b = \frac{2 \pm \sqrt{11}}{2}$

(j) $c = \frac{2}{3}$

(k) $y = -\frac{4}{3}, 2$

(l) $x = -3 \pm 3\sqrt{2}$

(m) $B = -5, \frac{7}{3}$

(n) $x = 2 \pm 2\sqrt{2}$

(o) Complex solutions

(p) $t = -6, 2$

(q) $x = \frac{3 \pm \sqrt{41}}{4}$

(r) $b = 5 \pm 2\sqrt{2}$

8.5: Answers

1.

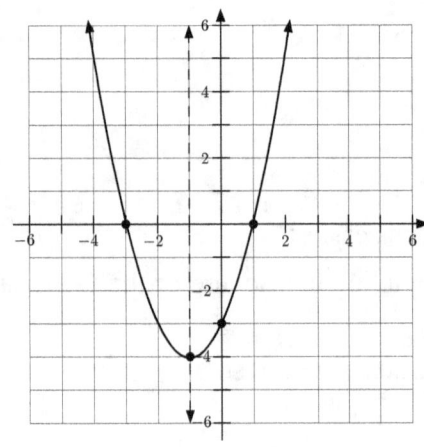

3.

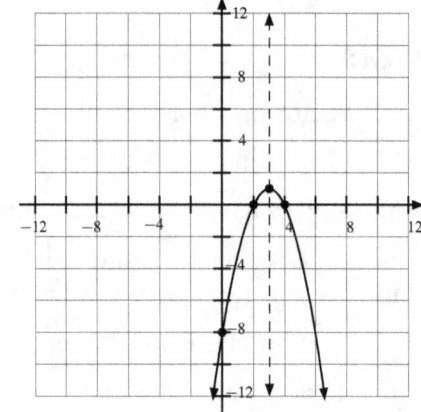

2.

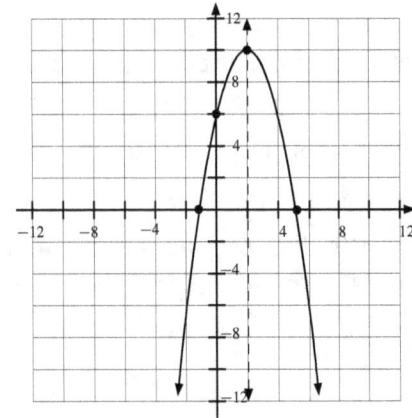

4.

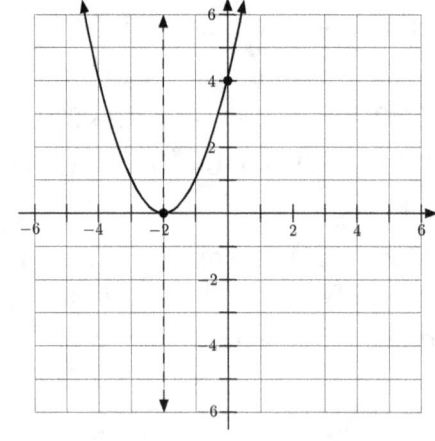

5.

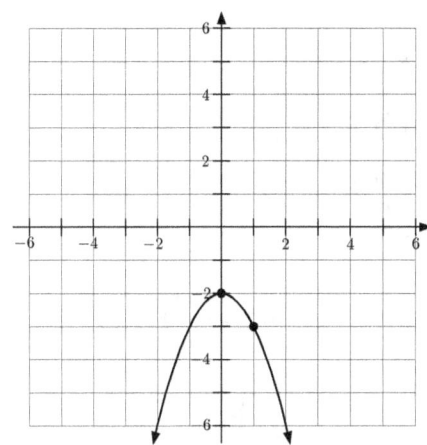

8.

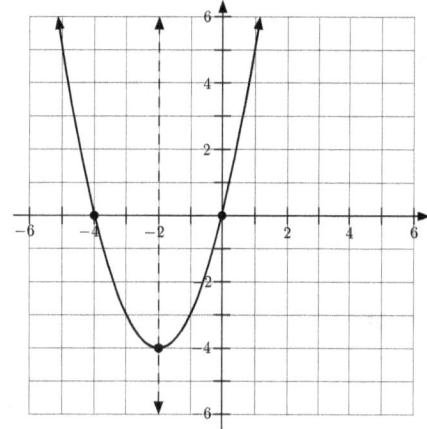

6.

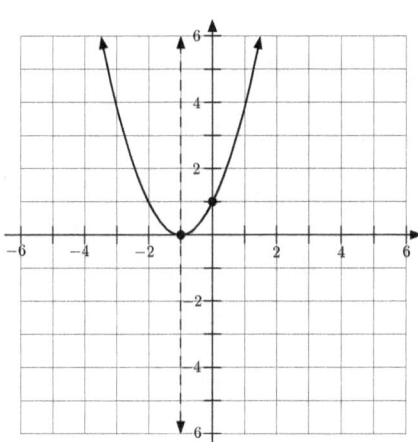

9.

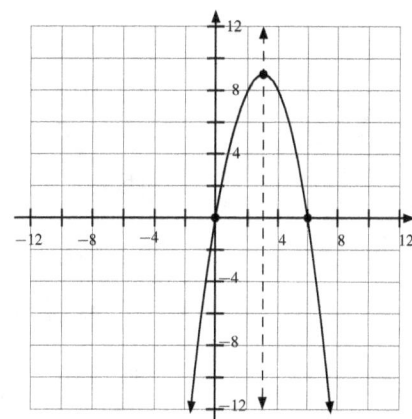

7.

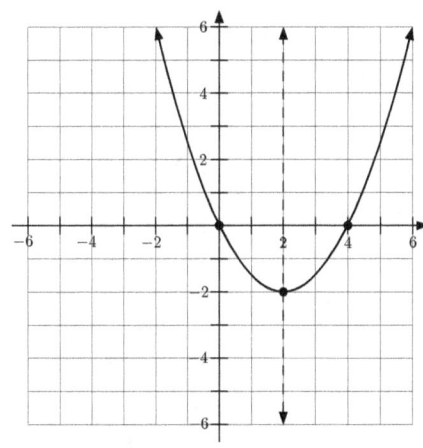

10.

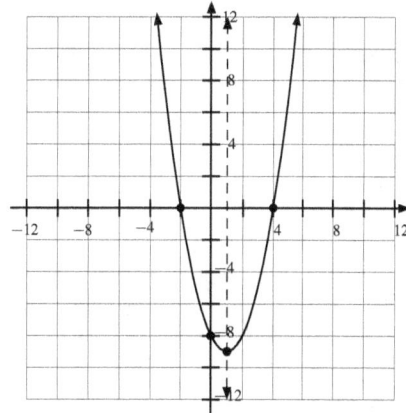

8.6: Answers

1. 43 inches

2. 3.2 and 6.2 inches

3. 6 yards by 12 yards

4. 2 inches

5. 6 yards by 9 yards

6. (a) $90\sqrt{2}$; 127.28 feet

(b) 64 ft/s; 43 mph

7. (a) 70 textbooks

(b) $775

8. (a) $888

(b) $2.70 or $3.00

(c) $974.70

(d) $2.85

9. (a) 144 feet

(b) 3 seconds

(c) 6 seconds

9.1: Answers

1. 6

2. Not Real

3. -5

4. $2\sqrt{7}$

5. $10\sqrt[3]{3}$

6. $7\sqrt{2}$

7. $-2\sqrt[3]{9}$

8. $21\sqrt{7}$

9. Not Real

10. $-21\sqrt{7}$

11. $10\sqrt[3]{2}$

12. $-10\sqrt[3]{2}$

13. $10\sqrt[3]{2}$

14. $\dfrac{\sqrt{2}}{2}$

15. $5\sqrt{3}$

16. $2\sqrt{6}$

17. $8|y|$

18. $10|n|\sqrt{n}$

19. $-10k^2$

20. $-4y$

21. $10h\sqrt[3]{h}$

22. $-3n\sqrt[3]{n^2}$

23. $-42|x|$

24. $4p^2\sqrt{2}$

25. $-5|g|\sqrt{2}$

26. $2|a|\sqrt{5a}$

27. $a\sqrt[3]{20}$

9.2: Answers

1. $5\sqrt{3}$

2. $-3\sqrt{7}$

3. $-3\sqrt[3]{6}$

4. $2\sqrt{7}-\sqrt[3]{7}$

5. $\sqrt{5}$

6. $-5\sqrt{6}-\sqrt{3}$

7. $\sqrt{10}-5\sqrt[3]{10}$

8. $4\sqrt{2}-\sqrt{5}$

9. $11\sqrt{6}+\sqrt{7}$

10. $-\sqrt{11}-3\sqrt[3]{11}$

11. $\sqrt{6}$

12. $5\sqrt{2}$

13. $11\sqrt{3}$

14. $-\sqrt{6}$

15. $-8\sqrt{2}$

16. $-9\sqrt{3}$

17. $-3\sqrt{2}$

18. $\sqrt{5}$

19. 0

20. $6\sqrt{2}+2\sqrt{3}$

21. $11\sqrt{6}-\sqrt{2}$

22. $\sqrt{10}-2\sqrt{5}$

9.3: Answers

1. $\sqrt{15}$

2. $-10\sqrt{2}$

3. $-36\sqrt[3]{3}$

4. $12\sqrt{30}$

5. $-12\sqrt[3]{2}$

6. $6|m|$

7. $10|r|\sqrt{r}$

8. $4|y|$

9. $2\sqrt{3}+2\sqrt{6}$

10. $5\sqrt{2}+2\sqrt{5}$

11. $15\sqrt{3}+15\sqrt{2}$

12. $10+8\sqrt{2}$

13. $4-\sqrt{3}$

14. 9

15. $25+4\sqrt{6}$

16. 5

17. $16-6\sqrt{7}$

18. $30+8\sqrt{3}+5\sqrt{15}+4\sqrt{5}$

19. 2

20. $4\sqrt{5}$

21. 20

22. $\frac{\sqrt{3}}{25}$

23. $\frac{\sqrt[3]{15}}{4}$

24. $\frac{\sqrt[3]{10}}{5}$

25. $\frac{2\sqrt{3}}{3}$

26. $\frac{\sqrt{15}}{3}$

27. $\frac{\sqrt{6}}{6}$

28. $\frac{\sqrt{10}}{15}$

29. Yes, it is a solution.

30. Yes, it is a solution.

31. Yes, it is a solution.

9.4: Answers

1. $x=125$

2. $y=-26$

3. No solution

4. $p=2$

5. $w=8$

6. $c=\frac{8}{3}$

7. $x=1$ or $x=5$

8. $x=\frac{1}{4}$ or $x=2$

9. $y=4$

10. $n=3$

11. $m=2$ or $m=-2$

12. $g=\frac{1}{4}$

13. $h=2$

14. $x=-1$ or $x=-2$

15. $k=-2$

16. $p=-1$

17. $m=3$

18. (a) $3\sqrt{5}\approx 6.71$

   (b) $2\sqrt{7}\approx 5.29$

19. $d=172$ feet

20. $(0,-2)$ and $(0,22)$

21. (a) $y=0$

   (b) $c=15$

22. Going from the 3rd line to the 4th line, Roody forgot about the x terms when multiplying $(x+2)^2$.

10.1: Answers

1. $\frac{3}{2}$

2. $-\frac{1}{2}$

3. $-\frac{1}{3}$

4. 0

5. undefined

6. 2

7. $k = -10$

8. $n = -\frac{1}{2}$

9. none

10. $m = 0$

11. none

12. $x = 1$

13. $p = 0$ or $p = 6$

14. $n = -3$ or $n = 3$

15. $a = -1$ or $a = 6$

16. $y = -\frac{5}{3}$ or $y = 2$

17. $\frac{1}{3n}$

18. $\frac{4}{x^2}$

19. 1

20. $\frac{a-7}{a+7}$

21. $\frac{10}{2p+1}$

22. $\frac{r-10}{r}$

23. 1

24. 8

25. $-\frac{1}{9}$

26. $\frac{1}{x+7}$

27. $\frac{x+7}{x}$

28. -1

29. $\frac{n+1}{6}$

30. $\frac{3a-10}{10+3a}$

31. $\frac{9}{v-10}$

32. $\frac{b+6}{b+7}$

33. $-\frac{k-4}{k+8}$ or $\frac{4-k}{k+8}$

34. $\frac{x-4}{3x-4}$

35. (a) 1/4

(b) The ys don't cancel.

10.2: Answers

1. $\frac{5}{56}$

2. $-\frac{39}{20}$

3. $\frac{22}{3}$

4. 18

5. -6

6. $\frac{1}{2}$

7. $\frac{81}{25}$

8. $-\frac{9}{7}$

9. $\frac{1}{8}$

10. $\frac{47}{16}$

11. $\frac{5}{2x}$

12. $\frac{5}{2(y-1)}$

13. $\frac{2}{z}$

14. $\frac{2(n+2)^2}{n}$

15. $\frac{4(w-4)}{w}$

16. $\frac{a+3}{3}$

17. $-\frac{2}{y}$

18. $\frac{-7(2c-9)}{3c(c+5)}$

19. $2h(7h+3)$

20. 1

21. $\frac{1}{2(v-10)}$

22. 1

23. $\frac{1}{5}$

24. -1

25. $\frac{1}{k+3}$

26. The -1 on the first line is incorrect.

27. The cancelling on the first line is incorrect.

10.3: Answers

1. $\frac{6}{a+3}$

2. $\frac{5}{y+2}$

3. $\frac{4}{x}$

4. 3

5. -1

6. $t+7$

7. $x-4$

8. $-\frac{1}{4}$

9. $\frac{43w}{20}$

10. $-\frac{3}{2y}$

11. $\frac{3x+4}{x^2}$

12. $\frac{2(3p-7)}{p^2}$

13. $\frac{5x+9}{24}$

14. $\frac{a+8}{4}$

15. $\frac{5r-12}{r(r-6)}$

16. $\frac{2(y-5)}{y(y+2)}$

17. $\frac{5}{x-5}$

18. $\frac{5m}{2-m}$ or $\frac{-5m}{m-2}$

19. $\frac{4-3w}{w+2}$

20. $\frac{5k^2+4}{k^2}$

21. $\frac{3(y+2)}{(y+3)^2}$

22. $\frac{4a}{(a+1)^2}$

23. $\frac{-n-1}{n(n+5)}$ or $-\frac{n+1}{n(n+5)}$

24. $\frac{2}{w-3}$

25. Roody can't just add denominators. The answer is $\frac{a+5}{5a}$.

26. The LCD is $x(x+3)$. The answer is $\frac{2x+3}{x(x+3)}$.

27. Roody dropped a "−" twice. The answer is still $x-4$.

10.4: Answers

1. $\frac{2x}{3}$

2. $\frac{x^3}{10}$

3. $\frac{5}{14x^2}$

4. $\frac{1}{18}$

5. $\frac{12+6x}{5x}$

6. $\frac{8-x}{32}$

7. $\frac{30x-3}{5}$

8. $\frac{15x}{6x-5}$

9. $\frac{18x}{9-x}$

10. $\frac{5x}{10x+6}$

11. $\frac{7x^2}{10x-3}$

12. $\frac{3x+6}{3x+2}$

13. $\frac{2x^2-2x}{3x^2+1}$

14. $\frac{x}{x-1}$

15. $\frac{3x^2+18x}{x+3}$

16. $\frac{2}{1-5x^2}$

17. $\frac{3x-7x^2}{21+14x}$

18. $\frac{y^2-27}{81}$

19. $\frac{8+a}{8-a}$

20. $\frac{c}{c-2}$

10.5: Answers

1. $x=6$

2. $m=-1$

3. $y=-\frac{4}{3}$

4. $z=\frac{11}{13}$

5. No solution

6. $y=2$

7. $x=\frac{1}{3}$

8. No solution

9. $c=-6$ or $c=6$

10. $y=\frac{1}{2}$ or $y=1$

11. $x=-3$ or $x=4$

12. $a=-2$ or $a=6$

13. $n=1$ or $n=-\frac{1}{2}$

14. $w=2$

15. $p=-5$ or $p=0$

16. $x = -2$ or $x = 9$

17. $a = \frac{1}{3}$

18. $c = -1$ or $c = \frac{3}{2}$

www.ingramcontent.com/pod-product-compliance
Lightning Source LLC
Chambersburg PA
CBHW081716220526
45468CB00008B/1867